# The Use of Econometric Models by Federal Regulatory Agencies

# The Use of Econometric Models by Federal Regulatory Agencies

**Joe L. Steele**
Texas Christian University

**Heath Lexington Books**
D. C. Heath and Company
Lexington, Massachusetts
Toronto          London

Published simultaneously in Canada.

Printed in the United States of America.

International Standard Book Number: 0–669–60962–5.

Library of Congress Catalog Card Number: 72–145901.

To Sandy, I.H.H.,
and H.H.L.

# Table of Contents

# List of Tables and Figures

# Foreword

Dean Joe Lee Steele's *The Use of Econometric Models by Federal Regulatory Agencies* constitutes a singularly appropriate addition to the list of titles in this series. The author is exceptionally well qualified to write this book. While writing it he held the First National Bank of Fort Worth Chair of Regional Economic Development and he is currently Dean of the M. J. Neeley School of Business of Texas Christian University. He serves also as Operations Research Advisor to the Dallas/Fort Worth Regional Airport Development Board, a major airport project, and as a consultant on transportation systems to the Battelle Memorial Institute. His teaching experience includes the field of economics at the University of Texas (Austin) and at Memphis State University and that of operations research at The American University in Washington, D.C. He has also been employed as an operations analyst by the General Dynamics Corporation and held the position of Senior Planning Economist with the National Aeronautics and Space Administration in Houston. In addition, he has worked as an Operations Analysis Advisor to the United States Air Force Chief of Staff. Thus, to this study, the author has brought not only the training and experience of an econometrician and operations analyst, but he has also brought that special training and experience provided by government service and practical application. Indeed, although he has utilized in his study only such research materials as are available in the public domain, his knowledge of the Washington scene has enabled him to utilize public documents and to have conversations which a less experienced and less sophisticated analyst would have missed. (For example, see his bibliography.)

Primarily, this book is concerned with an analysis and an evaluation of the first econometric model to be employed in a rate-making proceeding before a federal agency. Following this specific examination, an expanded discussion probes the scope and content of problems which may arise in the future use of such models before regulatory agencies in general. In his earliest chapters, Dean Steele not only examines the question of why an econometric model was first introduced in a case involving regulation of natural gas rates by the Federal Power Commission rather than in a proceeding involving some other regulatory agency, but he also discusses the concept of "an econometric model" in language comprehensible to a reader without mathematical training and provides a nonmathematical explanation of the specific model used in the *Permian Area Rate Cases*. He also discusses candidly the problems involved and procedures used in developing the model to fit that specific case.

As a counterpoint to this theme, his Chapter 3 includes a discussion and an analysis of the arguments concerning the relevance and validity of the model presented by the advocates of the positions of the opposing parties and the decisions made by the Hearing Examiner. The material in the first three chapters should be of special interest to students of government and business. The technical discussion of the model used in the *Permian Basin Area Rate Cases* in Chapters 4 and 5 should be of particular interest to econometricians, although readers without mathematical training can also benefit from reading it.

Dean Steele's technical critical evaluation of and suggestions for revisions of the model in Chapter 5 should also be of particular interest to econometricians, and the recommendations concerning the strategy for presentation and use of models in rate proceedings in general in Chapter 6 should be of great value to those who are preparing to formulate and present models in similar proceedings. One of the substantial contributions made by this book is the discussion on the relationship between a finding of fact and the components of a given econometric model. The question of the extent to which the use of such a model involves a finding of fact prior to an adjudication and decision by the Hearing Examiner is one which surely will not have occurred to econometricians and one which students of government and business and lawyers will find intriguing.

As Dean Steele points out, the functional relationships specified in an econometric model constitute assertions about facts, and the results of the statistical tests employed to estimate the values of parameters, that is, the estimated values of the parameters, constitute circumstantial evidence of a particularly sophisticated kind. A regulatory body may decide to adopt such findings of fact as its own; nevertheless, the duty of finding facts cannot be delegated by the Hearing Examiner or by the commission; the necessity that the decision be based on some finding of fact is clear.[1]

It does not follow, however, that a Hearing Examiner cannot, after a model has been put into evidence and successfully attacked, request that the model makers take account of any additional variables disclosed to be relevant by hostile testimony or of different data from those actually used. Such a procedure, as Dean Steele has noted, was in fact adopted in the *Southern Louisiana* proceeding. Arguments by lawyers steeped in the tradition of private trial practice that such a procedure involves an "impossible situation of having a witness change his testimony in the course of a trial" are irrelevant since they fail to recognize that the purpose of an administrative proceeding is resolution of conflicting social policy claims with the object of producing certain consequences deemed in the public interest rather than one of conflict between adversaries with incidental but inevitable social consequences.

In a rate-making proceeding, presumably the object of the regulatory agency is to fix the price at a level which will enable supply to keep pace with the growth of (or changes in) demand while not allowing the producers to capture excessive profits from their monopolistic positions. To the extent that the use of econometric models may help to clarify relationships and provide evidence of the significance of various factors, there can be no objection to development of the most appropriate and relevant model possible to serve that purpose. Indeed, Steele's book leads inevitably to a consideration of the question of whether or not the Federal Power Commission might be well advised to entertain a rule-making proceeding having for its purpose the development of an appropriate generalized econometric model which takes into consideration not only the ideas and concepts of the commission's staff but also those of the representatives of the industry, with the thought in mind of utilizing the general model developed in this way as a basis for employing specific models in particular future rate-making proceedings. In other words, specific content would be put into the general model on a case-by-case basis and frequent reviews of the general model could easily be conducted under the relatively relaxed conditions of a rule-making proceeding. Indeed, the Interstate Commerce Commission in 1966–1970 employed its rule-making procedure as a device for making a study of the cost estimating methods employed by its costfinding section (see ICC Docket 34013, October 10, 1966; and Press Release No. 156–70, July 30, 1970). I see no good reasons (other than legalistic ones) why a similar rule-making procedure cannot be adopted by regulatory commissions in developing general econometric models to apply in appropriate rate-making cases in instances in which groups of specific prospective cases can be classified as having similar characteristics.

This book is provocative and interesting; it well meets the criteria of books published in the Heath Lexington series: 1) it contains a strategy for social action; 2) it serves as a keystone for further investigation, as the reader can verify for himself by reading this little, but loaded, volume.

H. H. Liebhafsky
University of Texas

# Preface

This book examines the suitability of econometric models as primary tools of analysis for use by federal regulatory agencies. Since World War II, these agencies have been delegated greater responsibilities and at the same time have been asked to perform their functions with more precision and to take into account more fully all competing interests.

The procedural steps required for federal regulation in the modern United States economy are distinctly defined by law but the processes that compose each of these steps are not so clearly delineated. Regulation by federal agencies through the analytical method of an econometric model (or models) is a very real possibility for future commission decisions. The tool of econometrics is one that offers a system of sophisticated techniques powerful enough to sort out the complex forces that constitute the environment of a regulatory procedure and ultimately to place these forces in the framework of a descriptive mathematical model.

Major emphasis in the first portion of this book is given to the pioneering econometric efforts of the Federal Power Commission's Office of Economics. The introduction of an econometric model by an economist on the commission's staff in the 1961 *Permian Basin Area Rate Case* marked the first formal introduction into evidence of a complete econometric model in such a hearing. Chapters 2 and 3 in this book establish that the natural gas industry provided the Federal Power Commission with an almost perfect opportunity to develop and test the econometric approach to regulation. The Federal Power Commission's staff had been encouraged to seek new approaches to regulation by judges who had heard reviews of gas rate cases prior to the *Permian*. In addition to this legal exhortation to innovation, many economists who possessed an intimate knowledge of this industry have urged that the econometric approach be applied here.

The explanation and analysis of the Federal Power Commission's econometric models in the *Permian* and *Southern Louisiana* hearings form the foundation for this book's generalizations concerning the potential successful use of such models by other agencies. In the early chapters of the book I evaluate and criticize the processes of development employed by the Federal Power Commission's Office of Economics and the manner in which this type of testimony was introduced as evidence. The later chapters of this book are devoted to defining the alternative roles and scopes that econometric models may perform and encompass in regulatory cases. On the basis of this analysis an optimum staff presentation strategy is developed. Adoption of this presentation strategy would enable the competing interests to develop a useful econometric model in the course of the hearing, and such a model would protect the larger interests of all affected parties and provide a more precise vehicle of analysis to the hearing examiner.

The final analytical material covered in this book concerns the conflicts that may exist between the uses of econometric models by regulatory agencies and the constitutional guarantee of procedural due process of law. This vital constitutional issue may arise if econometric models are substantively used in future rate proceedings to supplant commission judgment. An econometric model fits into the issue of due process of law if it is an instrument used to "proportion" private rights and public interests. Due process of law demands that the regulatory process shall not be unreasonable, arbitrary, or capricious; and the manner of the use of an operational econometric model must be carefully investigated. If a commission decision in which a model has been employed is reviewed by a court, an interesting question may develop as to just how much of the given econometric model is to be considered a finding of fact.

As of now, no federal regulatory commission has actually used an econometric model as a primary vehicle for reaching its findings. The models of the Federal Power Commission's Office of Economics have made the most promising strides towards commission acceptance and ultimate use. This book's recommendations and conclusions as derived from Federal Power Commission experience are equally applicable to other regulatory agencies.

I am grateful for the help and encouragement I have received from a large number of people. Without question my most important source of aid was Professor H. H. Liebhafsky of the University of Texas, whose interest and expertise in economic theory and social control are well known to those who follow the current literature. If it were not for his impatient assistance these pages never would have required the services of a typesetter. I also received help from Professors Gerald Higgins and Howard Calkins of the University of Texas. Professor B. Joe Colwell of Southwestern University in Georgetown must be mentioned as a worthy contributor to this manuscript. Professor S. Allen Self of Texas Christian University gave me the benefit of his observations on my work and he provided several valuable suggestions for improvement. Professor Lewis C. Fay of the Texas Christian University Department of Journalism helped me immensely in the final preparation of this material. Of all my colleagues at Texas Christian University, I have received my greatest support from the late Dean Harrison of the M. J. Neeley School of Business.

Staff members of the Federal Power Commission have been most helpful to me in my research. Dr. Haskell P. Wald, who has joined the Staff as Chief of the Office of Economics since the *Permian* and *Southern Louisiana* hearings, has given me encouragement and aid. I want especially to thank Mrs. Phillis

H. Kline for her thoughtful criticism of my work while in manuscript form. Mrs. Kline's tenure with the Office of Economics traces back to the initial work on the Permian Model.

My wife, Sandy, typed all of the drafts of this book, and she provided many valuable suggestions for form and content.

Conclusions, criticisms and recommendations, of course, are my own responsibility—one I willingly accept.

# The Use of Econometric Models by Federal Regulatory Agencies

# 1 The Nature, Purpose, and Organization of this Study

The first use by a federal commission of an econometric model in a regulatory proceeding was in the *Permian Basin Area Rate Case*. In this proceeding, which was initiated in 1961, the Federal Power Commission's then-Chief Economist, Harold H. Wein, introduced a technique which was designed to ". . . provide the Commission with a flexible tool to deal with some of the inter-relationships between prices, supply, demand, and reserves, and the effect of other economic factors which influence the supply and demand for natural gas."[1] This "flexible tool" was an econometric model developed to project the impact that various ceiling prices would have on the demand for and supply of natural gas. It was also suggested in the presentation of this pioneering econometric study that such an analytical technique might be applied in future Federal Power Commission hearings and further that it might also be made a part of the "kit of tools" of other regulatory agencies involved with problems of rate making.

The importance, as a precedent, of the introduction into evidence of an econometric model in the *Permian* case is far greater than its actual influence on the decision indicates. The significance of the model's use lies in the attempt to establish by mathematics a structure that precisely described forces which constituted the economic environment critical to the rate proceeding at hand. Use of an econometric approach involves the tasks of defining and identifying primary economic forces, of making estimates of the magnitude of the influence each such force exerts on the dependent variable or variables under consideration, and of combining the results into a logical structure—the econometric model. In short, the econometric approach to regulation is an attempt to add new empirical dimensions to rate setting proceedings.

This study is much broader than an examination of the mathematical structure of a complex econometric model. The following chapters examine the total environment of regulation surrounding the first cases in which an econometric model was introduced. This material will have wide appeal to those who are interested in industrial economics, regulatory economics, economic history, and economic theory, as well as econometrics. For example, those trained in the more traditional approaches to the economics of industry and regulation will find that this work provides a gentle introduction to the purposes and methods of econometrics. On the other hand, the econometric

1

expert may find the discussions concerning institutional and legal intricacies to be as novel to him as are the quantitative techniques to the more conventional political economist.

This study has two purposes. The primary purpose is evaluation of the econometric approach to price regulation developed by the Federal Power Commission's Office of Economics in the *Permian* case. This evaluation will include an examination of the circumstances that led to the development of the original model; a review of the model's first application in the *Permian* proceeding and its concurrent use in the *Southern Louisiana* proceeding; a critical analysis of the model's basic structure; and a set of suggested model revisions which might make the model more effective. The secondary purpose of this study is estimation of the potential impact that the general use of a model of this sort might have in rate proceedings before other regulatory agencies. The study is organized as follows:

Chapter 2 discusses the broad scope of the regulatory authority of the Federal Power Commission and then examines in detail the events which led to the development of an econometric model to be applied to the regulation of natural gas rates. This chapter also considers why the econometric model was developed as an aid in a case involving natural gas regulation rather than in one involving any other industry, and more specifically, why the model was first introduced in the *Permian* case. Chapter 3 defines the concept of " an econometric model " as a general tool of analysis within the discipline of economics and then describes in detail the Federal Power Commission Model used in the *Permian* case. This first explanation of the model is presented in a nonmathematical context. Chapter 3 also provides a statement of the basic purpose of the model as developed by the Federal Power Commission's Office of Economics and a comprehensive review of the processes employed in developing the model to fit the requirements and meet the problems posed by the *Permian* case. Chapter 3 then examines the dimensions of the task of developing the specific econometric model used. The model was composed in fact of several components or " submodels." These submodels purport to isolate the major forces that determine economic relationships pertaining to natural gas prices. The primary submodels for supply factors on the one hand, and for demand factors on the other hand, were designed to permit independent estimation of alternative market equilibrium positions. The individual variables that are mathematically combined within these submodels were identified by the Federal Power Commission's Office of Economics as the most influential variables that could be used for the purpose of predicting changes in the market environment for natural gas. It will be suggested later that this identification constitutes a finding of fact. The issue of the model as a finding of fact will be considered in detail in Chapter 7.

Chapter 3 also examines each of the variables selected as essential to the model as well as some which were specifically rejected. In Chapter 7 it is established that rejection of a variable also constitutes a finding of fact. Finally, Chapter 3 contains an analysis of the positions taken in the *Permian* case by the Federal Power Commission's Office of Economics, as well as the decisions of the hearing examiner and the commission itself concerning the relevance and validity of the model as a basis for decision in that proceeding.

Chapter 4 is divided into two parts; the first section is devoted to a technical evaluation of the complete econometric model. This evaluation centers on the economic contents and on the statistical and mathematical processes embodied in the model. Chapter 4 also examines sources of data, assumptions, methodological approaches, and the mathematical application of the model with given input values. The second portion of Chapter 4 is devoted to criticisms of the construction and applicability of the model as employed in the *Permian* case.

Chapter 5 is devoted to proposals for revisions of the econometric model. The Federal Power Commission's Office of Economics has been working on what it describes as "an improved second generation model," to provide a more flexible and complete econometric tool for purposes of regulatory application. The general focus of the current Federal Power Commission revision is examined in this chapter, but the current revision is not yet complete and therefore cannot be critically reviewed. A suggested format for revision of the model is offered in Chapter 5; it was derived from the author's analysis of the original model and the sequence of relevant events since that model was first introduced. It is not intended to present a newly developed model that could serve as a tool in a rate hearing; our independent revision is concerned with suggested conceptual and methodological alterations in the existing model and not with the detailed development of a new operational model.

Chapter 6 explores the possible roles that an acceptable, operational econometric model might play in the regulatory process as it pertains to natural gas rates. This chapter also develops a suggested strategy for the most effective presentation and use of econometric models in rate proceedings. Consideration is given to potential degrees of commission reliance that could lawfully be placed on an econometric model—ranging from a minimum position in which the model is used only as a justification device to a position of maximum reliance on such a model, that is, to the case in which the model's estimates became the determinate guidelines for regulatory decisions without violating constitutional due process procedures. The emphasis in this chapter is on the wide middle ground of applicability of an econometric model (for gas price regulation) in a situation in which the model plays a supporting role in supplementing the customary procedures and analyses in rate hearings, but in which the ultimate duty of making the findings of fact is not delegated by the commission to its Office of Economics.

Chapter 7 contains an evaluation of the impact that use of the econometric model may have on the effectiveness of future natural gas rate hearings, particularly the way in which model building necessarily involves fact finding. This chapter defines the concept "due process of law" and details the procedural requirements of due process for regulatory commissions. The relationship between the constitutional guarantee of due process and an econometric model as a finding of fact is then considered, and definite conclusions are drawn from the analysis. The chapter concludes with a discussion concerning the problems involved in an extension of the econometric approach to rate proceedings carried on by other agencies in their specific areas of regulation.

Chapter 8 concludes the study. This final chapter contains a summary of the most important findings in the preceding chapters and a presentation of recommendations based on the data and analysis of the entire study.

# 2 Regulatory Background

The purpose of this chapter is to examine the broad scope of the regulatory authority of the Federal Power Commission over natural gas rates and to describe the sequence of events that led to the proposal for and development of an econometric approach for use as evidence in presenting the Federal Power Commission's staff's position in cases involving natural gas ceiling prices. This chapter also considers why such an econometric model was developed as an aid in a case involving natural gas regulation rather than in a case involving an alternative energy source, and, more specifically, why the model was first introduced in the *Permian* case.

The Federal Power Commission is an independent agency operating under the Federal Power Act (1920) and the Natural Gas Act (1938). This legislation gives the agency responsibility for regulation of the interstate operations of electric power and natural gas industries. The commission's authority includes, among other things, the power to issue licenses for the construction and operation of nonfederal hydroelectric power projects on government lands or on navigable waters of the United States, to regulate rates and other aspects of interstate wholesale transactions in electric power and natural gas, and to issue certificates for gas sales to and from interstate pipelines as well as certificates for construction and operation of pipeline facilities.[1]

Prior to 1938 there was no federal regulation of the natural gas industry. Limited regulation was undertaken for a brief period in 1919. In that year the Bureau of Natural Gas was established under the United States Fuel Administration. The Bureau of Natural Gas had no power to exercise control over prices; its function was to issue licenses to distributors and transporters in accordance with established World War I priorities. The Fuel Administration was deactivated after the conclusion of the war.[2] However, the fact that the federal government no longer exercised limited control over the natural gas industry did not leave that industry completely free from regulation. State governments in oil- and gas-producing regions had enacted numerous statutes prior to World War I to prohibit or control practices that it was believed led to excessive physical waste of natural gas. Such controls were largely looked upon as "conservation measures," and not as price control. However, an attempt by the state of Missouri in the early 1920s to prevent an increase in

5

the price of gas sold to local distributors by a company that brought the gas in from another state was declared in 1924 by the United States Supreme Court to be beyond the authority of state government on the basis that the Commerce Clause of the Constitution gave Congress the exclusive power to regulate interstate commerce.[3] This decision was made at a time, it may be noted, when the court was also interpreting the federal commerce power narrowly.

Thus, for 14 years after 1924 there was no regulation of interstate gas rates. Absence of effective federal regulation during this period does not mean that there was a complete lack of interest on the part of Congress concerning the level and structure of interstate gas rates. One of the more active agencies in this area was the Federal Trade Commission which, at the direction of the Senate, undertook an intensive investigation of the practices of the natural gas industry. The conclusions published as a result of this inquiry pointed to several industry abuses and helped to establish an undeniable need for remedial congressional action.[4] But it was not until 1938 that the Natural Gas Act was passed to "occupy the field in which the Supreme Court has held that the states may not act."[5]

In the 1938 act, the Federal Power Commission was given various new responsibilities as described below. Section 1(b) of the Natural Gas Act defined the jurisdiction of the commission in these words:

The provisions of this Act shall apply to the transportation of natural gas in interstate commerce, to the sale in interstate commerce of natural gas for resale for ultimate public consumption for domestic, commercial, industrial, or any other use, and to natural gas companies engaged in such transportation or sale, but shall not apply to any other transportation or sale of natural gas or to the local distribution of natural gas or to the facilities used for such distribution or to the production or gathering of natural gas.[6]

The commission initially did not believe that it had jurisdiction over the production and gathering of natural gas or over the field prices at which natural gas moving in interstate commerce was sold.[7]

Federal authority was expanded in 1942 by amendments to the Natural Gas Act which expressly gave the commission the power to attach to the issuance of a certificate of public convenience and necessity "such reasonable terms and conditions as the public convenience and necessity may require."[8] Certification of this type was required for interstate pipelines entering any market area. In 1942 a Circuit Court of Appeals decision expanded the Federal Power Commission's jurisdiction when it found that "sales of gas which has just moved interstate and sales of gas which is about to move interstate have like practical effects," implying that the law reached gas within a state.[9] The Supreme Court did not review this decision until 1947. The Federal Power Commission interpreted this decision as expanding its jurisdiction to include sale of gas by parties not affiliated with those who own interstate pipelines. Natural gas industry terminology for this interpretation was that the commission now claimed jurisdiction over gas sold "at arms length." "Sales at arms length" refers to gas sales by independent producers to parties who own interstate pipelines. The right of the commission to exercise jurisdictional control over independent producers was upheld by the Supreme Court in 1947.[10]

Independent gas producers reacted to this sequence of events by calling for legislation which would remove the commission's control over "arms length sales." Efforts by the independent producers to convince Congress were successful; in 1950, the Harris–Kerr Bill, which specifically exempted control over "arms length sales," was passed by Congress and sent to President Truman who vetoed the measure. In explaining this action, President Truman stated that he was "confident that the commission will apply standards properly suited to the special risks and circumstances of independent natural gas producers and gatherers."[11] This reaffirmation of jurisdiction over independent gas sales was followed by Federal Power Commission Order No. 154 in July 1950, which stated that the commission would investigate sales of individual producers or gatherers that had a material effect on interstate commerce if the rates appeared to be excessive.[12]

The current jurisdiction and responsibility of the Federal Power Commission were defined by the Supreme Court in 1954, in its opinion on the *Phillips Petroleum Company* case.[13] In 1948, the state of Wisconsin and the city of Detroit asked the Federal Power Commission to rule that the Phillips Petroleum Company was a natural gas company and that its gas sales price could thereby be regulated under the provisions of the Natural Gas Act. The commission ruled after investigation that the production and gathering activities of the Phillips Petroleum Company were not of the type which would bring the gas sales of this company under their control. This finding was made on the grounds that Phillips did not directly transmit gas interstate nor did it own any interest in transmission firms that piped gas interstate. The commission and Phillips both agreed that, although sales and subsequent movement of this company's natural gas were in interstate commerce, these sales were merely an incident of Philips' much larger interstate gathering business and not within the area of regulation.[14] The commission's opinion was based on an interpretation of the wording of Section 1 (b), cited above, of the Natural Gas Act which would exclude concerns that only produced and gathered natural gas. The commission's narrow interpretation of what constituted a regulable natural gas company was reversed by the Court of

Appeals.[15] The Supreme Court affirmed the Court of Appeals' reversal of the commission's ruling on this point, and held that the Natural Gas Act gave the Federal Power Commission a mandate under which it must set the price of all gas purchased for resale in interstate commerce.[16]

This decision, which established the current jurisdictional scope and responsibility of the commission, was not readily accepted by the friends of the independent gas producers. The 84th Congress in 1956 passed the Harris–Fulbright Bill, which expressly removed independent producers from utility-type regulation by the Federal Power Commission. This bill ostensibly would have imposed an approach to regulation of wellhead prices that stressed the inherent risks characterizing the production segment of the natural gas industry. The bill provided that fair rates of return be based on market value rather than on traditional cost-of-service calculations. This bill, which was sponsored largely by the oil and gas industry, passed the Senate with a 53 to 38 margin even though Senator Francis Case of South Dakota had announced during debate that he had been offered a direct bribe by a representative of the natural gas interests to support this bill.[17] Less than two weeks after the Harris–Fulbright Bill received congressional approval it was vetoed by President Eisenhower because of the open and illegal lobbying activity that had surrounded the bill's congressional progress. Despite his veto, President Eisenhower expressed agreement with the provisions of the bill that would give producers a more generous rate base and thereby a greater net return. The president was of the opinion that the promise of higher returns would encourage greater exploration and production of gas in the long run.[18]

Since the specific segments of an industry that are subject to price regulation have been identified, the next matter to be considered is the type of regulation to be imposed. Traditional regulatory rate-setting procedure is based on a cost-of-service approach. According to this method, a rate schedule is recommended at a level just sufficient to cover operating expenses, depreciation, taxes, and the maintenance of the "financial soundness" of the enterprise. The traditional cost-of-service approach has been applied by the Federal Power Commission in its regulatory role over interstate electric utilities,[19] but this approach has not been found to be as suitable for regulation of natural gas producers. In setting prices for natural gas production it is not feasible to use the cost-of-service approach because of the many difficulties in the measurement of traditional regulatory variables. One such difficulty involves the impossibility of assigning separate cost shares to natural gas and crude petroleum flows which come from the same field or from the same well. This problem of joint costs exists for products which cannot be separately produced, and any attempt to allocate total costs to arrive at an average cost for each of two or more joint products is futile.[20] Even if the problem of joint cost allocation could be overcome, a strict cost-of-service regulation procedure would provide rate structures that conceivably would differ from well to well in the same field, thereby rewarding producers who had spent relatively more on exploration with larger rate bases. A further difficulty in the use of the cost-of-service approach for natural gas producers arises from the administrative complexity of attempting to deal with the numerous producers on an individual basis.[21]

These regulatory problems did not exist prior to the Supreme Court's *Phillips* decision, which brought wellhead gas price regulation under commission control. This extension of the commission's jurisdiction introduced a new dimension into regulation that imposed problems not immediately recognized after the *Phillips* case. The first series of producers' petitions for rate increases were approved on the basis of cost evidence in the traditional manner. Initial producer requests for rate increases that were not supported by some sort of cost justification were uniformly denied.[22] On review of dismissals of early rate petitions not supported by cost evidence, the Court of Appeals for the Fifth Circuit approved the rulings of the commission. But in its approval, the Court of Appeals indicated that the commission was not bound to base its findings on conventional cost evidence if other evidence insured that consumers were not paying more than "just and reasonable prices," and that the commission need not fix separate rates for each producer if it was found that a uniform field rate was desirable.[23]

The Federal Power Commission's regulation of rates of electricity and natural gas under the Federal Power Act and the Natural Gas Act had always been on an individual cost-of-service basis prior to the Supreme Court's *Phillips* decision in 1954. This decision, which imposed upon the commission the responsibility of regulating thousands of gas producers' wellhead prices led to the adoption of an area method of regulation. Area rates are those determined by the commission to apply to independent producers' natural gas sales in each of 23 defined pricing areas. These areas are designed to combine all the producers in a common producing field into a common rate area.[24] The commission's Statement of General Policy No. 61–1 (1960) initiated the area rate approach by establishing guidelines for various producing areas. A summary of the intent of General Policy No. 61–1 was provided in the commission's Opinion No. 468 by the following:

The Commission in prescribing guideline prices for the various producing areas attempted to take into account all of the relevant information available to the Commission at the time. These guideline prices did not purport to be the just and reasonable rates, but rather the interim guides the Commission would follow in issuing certificates and in deciding whether to suspend rate increases until just and reasonable rates could be determined after hearings in area rate proceedings.[25]

Area rate pricing has been used by the commission since the issuance of General Policy No. 61–1 in 1960. The constitutionality of the area rate approach was upheld through the Supreme Court's decision in the second *Phillips* case and the *Permian* case.[26, 27]

In 1960, the commission met its responsibility for setting gas rates for sales in interstate commerce as established in the first *Phillips* case through the area pricing approach—a method which was approved in the second *Phillips* case. The area price guidelines first issued by the commission in 1960 were nothing more than interim prices for each producing area. When the commission issued these area price guides it was aware that this action was a temporary measure, which would have to be replaced by a system of area prices that could be defended as being "just and reasonable." "Just and reasonable" prices would have to be determined eventually, but in 1960, no acceptable or generally approved methodology was available and directly applicable to a problem of this type. As noted above, traditional public utility regulation methods were not applicable or suitable because of the impossibility of separating joint production costs, because of significant discovery cost variations in a single field, and because of the administrative difficulties of separate regulation of hundreds of individual producers. A new approach to regulation was required.

The remainder of this study examines one new approach to wellhead gas price regulation to meet the needs of the Federal Power Commission. This new approach employing an econometric model was developed by the Federal Power Commission's Office of Economics for application to the *Permian* proceeding. The order announcing the *Permian* proceeding was first issued in December 1960. Prehearing conferences were held in Texas in March 1961, and formal hearings were initiated in Washington in October 1961. Work on the *Permian* econometric model did not begin until May 1962.

This was the first area rate proceeding whose stated purpose was to determine just and reasonable rates for natural gas producers. The official proceeding spanned four and one-half years and resulted in Opinion No. 468 (issued August 1965) entitled *Opinion and Order Determining Just and Reasonable Rates for Natural Gas Producers in The Permian Basin*. The Federal Power Commission's Office of Economics staff's testimony focused on an econometric model which had been developed for this hearing. This method was introduced into the *Permian* proceedings in February 1963, to provide the commission with a flexible tool to deal with some of the most difficult problems of gas regulation. The hearing examiner and the commissioners, however, did not accept this specific econometric model as a significant or useful tool in the determination of the just and reasonable area rates for *Permian* Basin gas producers. The econometric approach as a method, however, was recognized by some of the commissioners as a promising one for use in area rate making. The basic econometric approach and the specific econometric model as introduced in February 1963, will be explained in detail in Chapter 3.

# 3 Introduction to the Permian Model

The purpose of this chapter is to define the concept of an econometric model as a general tool of analysis within the discipline of economics and to describe verbally, but in detail, the model used in the *Permian* case. This chapter also provides a statement of the basic purpose of this model—developed by the Federal Power Commission's Office of Economics—and contains a description of the processes that were undertaken to derive a model intended to meet the requirements of the *Permian* case. The description in this chapter of the econometric model includes an examination of the dimensions of the regulatory problem which constituted the environment of this model's development and an explanation of the major variables contained in the model. Finally, there is a brief review of statements made by the interested parties concerning the relevance of the model and its validity in that proceeding.

"Econometrics may be defined as the quantitative analysis of actual economic phenomena based on the concurrent development of theory and observation, related by appropriate methods of inference."[1]

Econometrics, as a methodology, attempts to combine a priori economic reasoning and empirical observations into a framework that will ultimately provide forecasts of economic behavior. Econometrics requires a combination of economic theory, mathematics, and statistics to provide an approach designed to permit the analysis to proceed from the highly abstract to the concrete. The success of an econometric model, used in this way, depends on: the "plausibility" of the economic theory that is being applied to the given real-world problem; the "proper" identification of the parameters that affect the economic behavior under analysis; the development of a "viable" model structure to support the selected parameters; and the derivation of accurate coefficients. Econometric models, essentially, are mechanisms of concise description of selected observed economic phenomena. The interest of the econometrician, however, is usually not in mere description—it is in forecasting. Econometric models are primarily developed as tools for forecasting economic behavior, but these forecasts are subject to error due to the inherent difficulties of econometric model building. Errors in prediction may be a result of the evolutionary changes in parameters brought about by fundamental alterations in technology or institutions; and errors may also be traced to inaccurate parameter estimation resulting from statistical problems of identification, bias, serial correlation of errors, and other such difficulties.

11

Even though the econometric method is difficult to apply and although its forecasts are subject to error, in recent years significant numbers of economists have become convinced that improvements in economic policy must be based on econometrics.[2] In introducing the research strategy for and application of the *Brookings Quarterly Econometric Model of the United States*, Duesenberry and Klein noted that:

Only a few econometric models have been used for forecasting over an extended period. In spite of all the difficulties of forecasting they have produced forecasts substantially better than naive forecasts and substantially better than those of most ad hoc forecasters. But the case for the usefulness of econometric models in forecasting does not rest entirely, or perhaps even mainly, on the record of past success. Naive forecast methods are, almost by definition, difficult to improve. Econometric models can be improved. . . . But no one, and least of all the makers of the present models, will doubt that there are real possibilities of improving the performance of existing models.[3]

Econometrics therefore is a method designed for those whose concern with the quantitative and quantifiable formulation of current economic situations is primarily focused on prediction. This methodology is highly suited to considerations of strategy and policy, and even though a carefully developed econometric model may not guarantee highly accurate forecasts, the standard processes that are undertaken in developing an econometric model may probe more systematically than other approaches to economic analysis. Indeed, the real value derived from the use of an econometric model may lie in the novel questions produced and not in its selective tentative solutions. A major virtue of this approach lies in the fact that its final product is an explicit econometric model. Such a completed model is tangible and subject to statistical testing and evaluation by others who are parties to an adversary proceeding in which the model has been introduced as evidence. Because use of an econometric model involves a precise formulation of the issues involved, other analysts can manipulate the model to investigate how selected changes in model assumptions, parameters, structure, or coefficients affect the variables dealt with by the model. The ability of other analysts to manipulate a model removes much of the aura of mystery that surrounds many of the more naive predictive methods and provides an almost ideal method of crystallizing the issues in an adversary rate proceeding.

One of the terms mentioned several times in the above exposition requires special attention. That term is "identification." In the language of econometrics, identification relates to the unique estimation of the model's parameters. The typical econometric model is composed of a system of simultaneous equations developed through a combination of a priori information and statistical analysis of empirical data. The a priori information concerns the theoretical economic forces the econometrician employs to establish the basis for the structure of his system. The statistical techniques applied to the real-world data attempt to estimate precise values for equation variables. The discovery of historical correlations among variables constitutes the principal finding from such statistical analysis. The identification problem centers about the difficulty of assigning precise degrees of influence associated with structural variables in a given model. For example, if economic theory states that the price of a given good is determined by the interaction of those factors that define supply and demand and that price is a single data point at a given moment in time, how can the econometrician identify the separate schedules for supply and demand? Statistical procedures have been developed to assist in the identification process; however, most economic processes have unique characteristics which require some structural simplifications to permit identification. Therefore, the demand curve for a given product may be identified from price data if a convincing argument is made that, for a given time span, the factors influencing demand were relatively stable while those affecting supply were not.

The Federal Power Commission's Office of Economics developed an econometric model for the *Permian* hearing for the purpose of projecting the effects that a selected range of wellhead gas prices would have on supply and demand for natural gas. This model, as conceived, used as inputs an assortment of hypothetical gas prices and resulted in the production of estimates of future exploratory effort, of additions to gas reserves, of gas production, of gas consumption, and of other factors which might affect the natural gas industry and the public interest. In initiating work on this model, the Office of Economics demonstrated its awareness of the full meaning of the observation by Justice Brandeis that: "In no other field of public service regulation is the controlling body confronted with factors so baffling as in the natural gas industry."[4]

The development of the econometric model was an attempt to crystallize some of the economic issues involved in regulation. It was designed to serve the interests of the natural gas industry and those of the consuming public, and was intended to serve a regulatory function such as that described by Justice Jackson in the *Colorado Interstate* case:

Far sighted gas regulation ... will use price as a tool to bring goods to market ... to obtain for the public the needed amount of gas. Once a price is reached that will do that, there is no legal or economic reason to go higher; and any rate above one that will perform this function is unwarranted .... On the other hand if the price is not a sufficient incentive to bring forth the quantity needed ... the price is unwisely low. The problem, of course, is to know what price level will be adequate to perform this economic function.[5]

Justice Jackson's statement of the proper role of gas regulation would appear to rule out regulatory methods that do not take into account the entire complex environment of the natural gas market in which gas prices are determined. This regulatory procedure would consider such factors as exploration, production, collection, distribution, and final burner-tip consumption. Regulation, which attempts to harmonize the conflicting individual and collective interests expressed in this market, is a process which involves broad planning and ultimately requires the development of a national policy with respect to power sources, at least to the extent that such regulation purports to be a matter of national interest.

In constructing its econometric model, the Office of Economics did not attempt to solve all the regulatory problems of the gas industry; nor did the model builders intend to construct a device that would set the "right price" for any specific area. Such a model should be developed to support what may be called the strategy of regulation. That is, this type of model ought to be designed to be used repeatedly in a process of trial and error rejection as new data are developed.[6] The contribution envisioned for the econometric model was that of providing a technique that would contribute to the analytical process of arriving at the right price, and, at the same time, establish a firmer basis for final decision.

The *Permian Basin* model was developed in three separate parts. Separate analyses were undertaken of exploration activities, of residential and commercial consumption, and of industrial consumption. These three broad categories were judged by the Office of Economics to be the most significant factors in this area rate proceeding. The basic analytical approach used for developing each of these three separate components was regression analysis. Regression is a statistical process that provides a technique of estimating the value of one dependent variable corresponding to that of a given value of a selected independent variable. This process is accomplished by estimating the value of the dependent variable from a least squares curve fitted to the data. (The least squares statistical method involves minimizing the squared deviations of actual observed values from a line of "best fit" drawn through the data, and hence, defines the equation of the curve.) Regression analysis is a technique of estimation which attempts to measure the degree of the relationship existing between a dependent and an independent variable.

Regression analysis has been developed to deal with problems involving a requirement of estimating the value of a dependent variable. Obviously the process involves the selection of the proper variable or variables believed to vary systematically with respect to changes in the series to be estimated. In the *Permian* case, the Office of Economics selected three major dependent variables as the nuclei of their model. As noted above, these dependent variables were exploration activity, residential and commercial consumption, and industrial consumption. The behavior of each of these dependent variables was analyzed separately by means of multiple regression analysis. A brief descriptive explanation of the derivation of the estimating relationships that comprise the major framework of the *Permian* econometric model follows.

The first problem resolved in the submodel developed to estimate exploratory activity was the selection of an index used to represent the unit of exploration. Exploratory activity had been measured in other natural gas industry studies by such factors as the number of geological, geophysical, and geochemical crews employed; the amount spent on exploration; the total footage of exploratory drilling; the number of wildcat wells drilled; and other similar items. The staff selected the number of exploratory wells drilled each year as the best measure of exploratory activity. The American Association of Petroleum Geologists defines the term "exploratory wells" as one that includes the following drilling activity: "outpost" wells, which are attempted deep extensions of a partly developed pool; "new pool" wildcats, which are wells outside the limits of a proven area; "deeper pool test wells," which are drilled within the limits of a proved area but below the deepest producing pool; "shallow pool test wells," which are drilled within the limits of a proved area but above known pools; and "new field wildcat wells" which are drilled on a structural environment with no previous productive history. The Office of Economics initially considered using drilled footage as the index for exploratory activity, but the number of exploratory wells drilled was chosen as the best measure of exploratory activity because it was believed that the unit of decision is essentially a decision to drill a well and not necessarily to drill a foot or some number of feet.[7]

Once the definition of the unit of exploratory activity had been determined, the next step was that of determining the influence exerted on this variable by various independent variables. The first independent variable selected as having a strong influence on exploratory activity was the expected future price of the products to be produced. Drilling activity may produce gas, liquids, or a combination of both; therefore, the price variable chosen was a combination of the current prices of each. It was assumed that current prices would form the best estimate of future prices. Thus, for the variable of expected price of all liquids the current crude oil price was employed; and for the expected prices of gas, a standard price index known as the Foster Associates series for the "Initial Base Price of Gas" was used. This "Initial Base Price of Gas" is a measure of the price as recorded in new gas producer–gatherer contracts in the major oil and gas producing states. This price series was selected in preference to the use of a simple average of all current field prices because current field prices include some price agreements made under contracts negotiated in previous years. All price data used in the exploratory activity submodel were adjusted by the U.S. Department of Labor's wholesale price index.

Another independent variable selected for investigation in this submodel was production of oil and gas. This variable was chosen to constitute a proxy for an aggregate producer revenue series. The Office of Economics was not able to develop a satisfactory revenue series that would distinguish liquid and gas revenues, and therefore, it employed the production series.[9]

The third independent variable selected for inclusion in the submodel was the cost of money as measured by the average yearly interest rate in major cities for commercial loans. This variable was employed only in the initial experimental applications of the submodel and was later eliminated as statistically insignificant. The early inclusion of a cost-of-money variable was a result of an a priori judgment that exploration activity would rise and fall with the cost of obtaining capital; this conclusion was later judged to be incorrect.

The fourth variable in the submodel was an "exploration success" ratio which related the number of successful exploratory wells to total exploratory wells drilled, as reported by the American Association of Petroleum Geologists. This variable was included because it was reasoned that more exploratory activity would take place when the prospects of drilling success were favorable, and that exploratory activity would be reduced when the success ratio was declining. This variable was first introduced into the submodel on a current year basis and then as a one-year lagged series. There was no apparent statistical difference in the inclusion of the current versus the lagged data, and the current-year success ratio was arbitrarily selected.

The fifth independent variable seriously considered in the submodel was an index which would measure the probable depths of new producing reservoirs. The average depth of exploratory wells has been increasing with time, and deeper wells are proportionately more expensive to drill per foot than shallow wells. This variable was included because of the judgment that as the probable depth of new producing reservoirs increases, the number of exploratory wells should decrease. The final independent variable included in the exploratory activity submodel was an aggregation of factors which could not be separately identified by the Office of Economics. This catchall category was labeled "influences varying systematically with time."

The exploratory activity submodel therefore related the dependent variable, number of exploratory wells drilled per year, with the independent variables of crude oil and natural gas prices, crude oil and natural gas production, the prospects of drilling success, the probable depths of new producing reservoirs, and other influences varying statistically with time. These variables were identified for the twenty-two year period from 1940 through 1961, and these data were then employed in a multiple regression analysis to produce the equation that formed the exploratory activity submodel. This submodel was the source of the formula that the Federal Power Commission's Office of Economics derived to relate various values of the independent variables to levels of exploration. This submodel was capable of being used to estimate future levels of exploration for a variety of hypothetical values for each of the other variables. For example, the submodel was used to forecast five separate levels of exploration corresponding to five wellhead gas prices. To reduce the amount of calculation which would be required for computing values for all ranges of each of the variables, selected point values were derived and curves were plotted to provide estimates for intrainterval values thereby providing shorthand graphical estimates. The equation for this submodel as developed from the described time series data will be presented and evaluated in detail in Chapter 4. A description of the two demand submodels follows.

The residential–commercial submodel was developed through an analysis of selected variables which were thought to determine demand for gas in each of the 44 states in which gas was available for residential–commercial consumption. The industrial demand submodel was developed from an investigation of gas demand factors in the 42 states that had an industrial supply of gas during the period of analysis. These demand submodels were constructed from data gathered on a cross-section basis in contrast to the time series approach used for exploratory activity. Therefore a single year was used for the base in the two demand submodels and observations from each state comprised the data that were subjected to a multiple regression analysis to provide the desired estimating relationships. The independent variables selected in the residential–commercial analysis were the burner-tip price of natural gas, prices of competing fuels, weather factors, and the number of residential and commercial customers served in each state by natural gas. The unit of measure selected for the burner-tip price was the average current price per Thousand cubic feet of gas used by residential and commercial customers. For a measure of the prices of competing fuels, the average delivered price of light fuel oil was selected. Coal prices and electric rates were considered here, but were rejected as being insignificant competitors of gas in residential–commercial markets. The weather factor variable was represented by a weighted degree-days series in which a degree day is defined as a unit of measurement equal to one degree of variation from a standard temperature. The unit of measure for the final variable in the residential–commercial demand submodel was simply the number count of gas customers served in each state. The values for these variables were determined in each of the states for the years 1955 through 1961, and a separate estimating function was developed for each year through multiple regression. The resulting functions for each year were found to be quite similar, and the 1961 function was selected because it contained the most recent data.

The industrial demand submodel was developed through the same cross-section multiple regression approach as was the residential–commercial function. The industrial demand submodel was derived from identification of six independent variables which were as follows: the industrial gas price as defined by the average burner-tip price paid by industrial consumers, the industrial oil price, the industrial coal price, industrial employment, the number of kilowatt hours generated by steam, and the number of degree days. Oil and coal prices were included in this submodel because both compete with natural gas for industrial use. Industrial employment was included to serve as a proxy variable for measuring the industrial output of each state. Kilowatt hours generated by steam were included in this list of variables because steam generating plants comprise a major industrial market for natural gas. Values for these variables were identified for each of the states for the years 1955 through 1961 and were used to develop estimating functions through multiple regression analyses for each year. As in the residential–commercial submodel, the 1961 function was selected as the proper forecasting tool.

The Federal Power Commission's Office of Economics introduced into evidence its econometric model in the *Permian* hearing on February 20, 1963. This model—which was designed to predict future schedules of supply and demand based on selected values for given independent variables—was submitted by the chief economist of the Federal Power Commission, and Federal Power Commission expert witness, Harold H. Wein. In his introduction, Dr. Wein argued that this model would deal effectively with some of the most difficult problems of gas regulation, and that it would provide a means whereby the commission could keep in touch with changes in the natural gas industry and adjust its regulatory policy from time to time as one or another of the independent variables that determine exploration and consumption depart from the assumptions upon which the projections were based.

The econometric model that Dr. Wein introduced is considered in mathematical detail in the next chapter. The broad conclusion of this model is summarized here to introduce the structure of observations drawn by the commission's econometric expert. The most basic finding reported was that gas price increases tend to reduce the volume of gas sold. The supply of gas is essentially the underground inventory of proven gas reserves. Inventory levels are referred to as "life indexes" in which proven gas reserves are divided by annual gas consumption to determine how many years gas supply is available at any given time. Dr. Wein's econometric testimony suggested that with reduced consumption (following a price increase) the life index value would rise, meaning that more years of gas supply would be available without the addition of any new proven reserves. The conclusion, therefore, was that price increases do not stimulate new exploration, and, in fact, price increases tend to reduce the required volume of gas held in reserve as inventory. On its surface, this line of argument seems to be counter to conventional economic theory which usually associates commodity price increases with increased offerings. The commission's econometric witness commented on this point by stating:

The results may appear to be somewhat paradoxical at first glance. But things are not always what they seem. And this is particularly true in economics where it is deceptively easy to give simple answers to questions which seem simple, but are really complex. Thus one finding which we have obtained, as a result of comprehensive analysis is that an increase in wellhead prices of new contracts has not (after all other relevant factors have been accounted for) resulted in additional new gas reserves (except, perhaps, temporarily). If the past is any guide to the future, price increases will lessen exploration, and hence result in the long run, in less gross additions of new reserves. This seems curious to those who believe that increased price will always result in more supply no matter what. But another proposition commonly held, and which is true, is that demand will fall as prices rise, other things equal. And if demand falls as prices rise, we will either have excess capacity (i.e., excess stocks of gas) if businessmen keep on adding to supply, as the first proposition holds, or they will be prudent and cut supply to meet the demand.[9]

A brief account of selected statements concerning the degree of acceptance by interested parties of the econometric model will help to establish a perspective for the role that the model assumed in the first area rate proceeding to determine just and reasonable prices. The assessment of the model by the oil and gas producers was summed up in their "Joint Initial Brief" as follows:

The Wein Study was prepared by the witness and a number of other staff personnel. The mechanically derived results reflect their absence of understanding of the implications of the study. . . . Even if the Wein Study were free from inaccuracies and subjective manipulations . . . , it would not be a useful tool in regulation.[10]

It will be noted in following chapters that the respondents in this hearing did not attempt to refute the validity of the econometric method as a tool for policy; their attack was focused on the specific model as introduced by Dr. Wein.

In hearing examiner Wenner's initial decision, the econometric model is mentioned only occasionally, and it is never referred to as an instrument that directly influenced his findings. He dismissed the applicability of the Wein model by stating that "the econometric study presented here is not relevant or material to the problem."[11] "The Opinion and Order Determining Just and Reasonable Rates for Natural Gas Producers in the Permian Basin," which was issued as Opinion No. 468 by the Federal Power Commission, did not favorably mention the econometric model.

The econometric model thus assumed a prominent role only in the early phases of the *Permian* proceeding when roughly one-fourth of the staff's brief to the examiner was devoted to that study and more than 1,800 pages of transcript were required for model testimony, rebuttal, and cross-examinations. After extensive consideration, the Wein model was judged by the examiner to be inapplicable to the regulatory problem at hand.

It should not be assumed, however, that the use of econometric techniques in the regulatory procedures of natural gas would depend solely on the success of this first model. The initial promise of the Wein model was not substantiated, but the criticisms leveled at this study were against specific weaknesses of the staff's model and not against the concept that a model could be developed that would perform a meaningful supporting function in the regulatory process.

Leaders in the executive, legislative, and judicial branches of government had been calling for a new, improved approach to the regulation of natural gas; and in response, the econometric method here discussed was devised. The pioneering model may not have accomplished its objectives, but it did open a new dimension for inquiry in the regulatory proceedings. Chapter 4 will provide a detailed evaluation of the Wein model as presented to the hearing examiner in February 1963.

# 4

# Evaluation
# of the Permian
# Model

The purpose of this chapter is to provide a thorough technical description and evaluation of the specific econometric model developed by the Federal Power Commission staff for the *Permian* proceedings. In the pages that follow, the economic, statistical, and mathematical processes embodied in the model are examined. Particular attention is devoted to data, assumptions, methodology, and mathematical applications. The chapter concludes with a general criticism of the construction and applicability of the model as employed in the *Permian* case.

The preceding chapter has broadly examined the nature of the econometric approach, and generally described the format of the *Permian* econometric model. Complete comprehension of this model, however, requires a detailed delineation of each of its structural components. Chapter 3 also introduced the exploratory activities submodel and the two submodels which were developed to estimate residential–commercial and industrial demands for natural gas. The exploratory activity submodel used to estimate parametric values for natural gas supply is joined to the demand submodels by a "linkage submodel." This linkage submodel integrates the other submodels into a single united model by permitting the same basic price value to be used as a common input. The model is composed of eleven mathematical equations organized as illustrated by the flow chart in Figure 4–1. This flow chart is arranged so that the components of the exploratory activities submodel are arrayed across the upper section; the linkage submodel occupies the dotted blocks at the lower left; and the demand submodels are shown as the remaining lower blocks. This flow chart includes all of the basic equations that comprise the model and illustrates the sequential processes involved as one moves through the model from the initial input of the basic price of gas to the estimations of total national consumption and marketed production of natural gas in a given time period. Figure 4–1 provides our first complete illustration of the model which has been previously discussed in only very general terms. The appendix to this chapter includes an even more detailed presentation of the equations and a thorough discussion of the variables.

Chapter 3 further explained that an econometric model is essentially a formal mathematical representation of the notions that the model builder has about a phenomenon.[1] The flow chart in Figure 4–1 explicitly details the

21

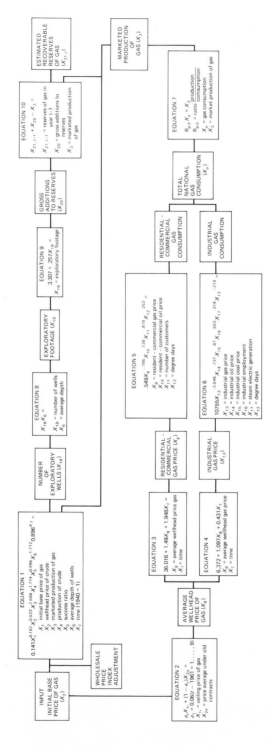

**Figure 4-1.** Flow Chart *Permian* Rate Hearing Econometric Model, November 1963.

notions and hypotheses the Office of Economics had about natural gas explorativity activity, production, and prices during the *Permian* hearing. If a model is to constitute the formal representation of such notions, then that model should possess the element of "reasonableness." This basic reasonableness may be tested by evaluating the appropriateness of the variables selected for each equation and by evaluating the coefficients assigned to the variables; further, evaluation may extend to examining the reasonableness of the manner in which the equations are linked to form the model.

How reasonable was the *Permian* model? In the broadest sense, those responsible for the model recognized that since the Federal Power Commission was responsible for fixing prices at the wellhead for natural gas, the model would be required to accept candidate values for the price of gas as inputs and to estimate for each price the corresponding volume of gas demanded and supplied. Those who designed the model wanted to develop a mechanism that would forecast effects resulting from changes in the wellhead price of gas on marketed production of gas and on new exploratory activity. The purpose of the model was thus clearly *reasonable*, but for a more critical appraisal of its reasonableness the most important equations must be examined in detail. Before beginning an evaluation of individual equations, however, a comment should be inserted about the meaning of the term "reasonableness" in this context. It must be asserted that a reasonable approach does not mean "the correct" approach because there is no way to define in detail "the correct" econometric approach toward the solution to a specific regulatory problem. All would agree that a reasonable approach should not be seriously inconsistent with generally accepted and applicable economic theory or mathematical and statistical technique. Different models that deal with the same problem may have been developed through approaches that were reasonable in the above sense, but these models may point to radically different solutions to a common problem. Therefore, to state that a given model is reasonable does not necessarily imply that it is a "good" guide for policy.

A useful purpose may be served by referring to a related econometric investigation prior to the detailed analysis of the *Permian* model. Franklin M. Fisher's *Supply and Costs in the U.S. Petroleum Industry* contains two econometric studies which deal with those supply aspects of petroleum as determined by wildcat drilling ventures. Fisher's studies focus on crude petroleum and he does not attempt to separate oil statistically from gas exploration ventures. He explicitly states in his investigation the following:

No distinction is made between gas and oil wildcats, as that classification is arbitrary and entirely a matter of hindsight. Since oil and gas are joint products and the search for one is inseparable from that for the other, they are treated symmetrically.[2]

The Fisher analysis suggests that price increases for crude oil lead to more wildcat drilling. New wildcat wells will be drilled on higher risk locations that have been made more attractive by the crude price increase. These locations had previously been judged unworthy for exploration at the lower price. As a result of a given increase in price, Fisher explains, one would expect to find a more than proportional increase in wildcat drilling activity, but since the new exploration takes place on higher risk locations the proportion of new discoveries will decrease.[a]

The Fisher model is interesting in this context even though it is not directly comparable to the Wein model. Fisher's focus on supply indicates the relative insensitivity of new crude discoveries to price increases, and Wein's analysis of demand suggests a high degree of sensitivity for an associated reduction in volume of gas demanded, given a price increase. The material below considers, in detail, the composition and the properties of the Wein *Permian* model.

Equation 1 is the most essential equation in the exploratory activities submodel (as illustrated in Figure 4–1) because it relates the dependent variable—number of exploratory wells—to the relevant independent variables (listed within the box of equation 1). The independent variables within this equation were not all representative of the staff's conceptual notions of what the variables should be. Table 4–1 indicates the choices of conceptual variables in comparison to those actually employed. Some of the conceptual variables were not used because of a lack of data, but conceptual variable (5) " cost of obtaining capital for exploration " and statistical variable (5) " interest rate " were omitted from the model because they were judged to be statistically insignificant. Equation 1 in the *Permian* model was:

$$Y = 0.141 X_1^{0.162} \ X_2^{0.522} \ X_3^{1.566} \ X_4^{1.774} \ X_5^{0.996} \ X_6^{-1.712} \ 0.896^{X_7}$$

where: $Y$ = number of exploratory wells (in thousands)

$X_1$ = initial base price of gas (¢/Mcf) (i.e., ¢/Thousand cubic feet)

$X_2$ = wellhead prices of crude ($/bbl.)

$X_3$ = marketed production of gas (billions of Mcf)

$X_4$ = production of crude (billions of barrels)

$X_5$ = success ratio (percent of productive well to total)

$X_6$ = average depth of exploratory wells (thousands of feet)

$X_7$ = time (1940 = 1, 1941 = 2, etc.)

---

[a] Fisher's model, which was derived from 1946–1955 data, related a 1 percent increase in the price of wellhead crude with a 2.58 percent increase in wildcat drilling. The associated success ratio for the new wildcat wells drilled on marginal locations was reduced to about 0.36 percent, and the average size of discovery declined by about 2.18 percent. Therefore, a 1 percent price increase would set in motion a chain of events which would increase new discoveries by about 0.31 percent.

**Table 4–1. Exploratory Model—Based on 1940–1961 Experience (Dependent Variable—Number of Exploratory Wells) Independent Variables**

| Conceptual Variable | Closest Obtainable Statistics |
| --- | --- |
| (1) Expected natural gas price | Initial base price of gas |
| (2) Expected crude oil price | Current wellmouth price of crude |
| (3) Marketed gas production (in Mcf) | Marketed gas production |
| (4) Marketed production of crude liquid (in bbls). | Crude production |
| (5) Cost of obtaining capital for exploration | Interest rate |
| (6) Prospects of success | Exploration success ratio |
| (7) The probable depth of new producing Reservoirs | Average depth per exploratory well |
| (8) Other influences varying systematically with time | Calendar time |

*Source:* Harold H. Wein, *Natural Gas Supply and Demand* (Washington, D.C.: Federal Power Commission, Docket No. AR 61–1, 1963), p. 31.

This equation expresses the dependent variable (number of exploratory wells) as the product of exponential functions of the independent variables, and the exponents of the independent variables may be interpreted as "elasticities." The elasticity of a dependent variable is the ratio of the proportional change in the dependent variable corresponding to the proportional change in the independent variable. For equations in the exponential form of equation 1, the elasticities are the same at all points which relate a dependent variable to a given independent variable as long as the other independent variables are held constant. For example, variable $X_1$ (the initial base price of gas) has an exponent of 0.162 which may be interpreted to mean that if the initial base price of gas is increased by 1 percent, $Y$ (the number of exploratory wells drilled) will increase by 0.162 percent, provided that all the other independent variables are held constant. The exponent of $X_4$, 1.774, indicates that a 1 percent increase in crude production will be associated with a 1.774 percent increase in the number of exploratory wells. These numerical exponents may be loosely interpreted as measurements of the sensitivities of the dependent variables to changes in their respective independent variables while the values of the other independent variables are held constant. The estimated values of the exponents—regression coefficients—may be positive or negative, indicating whether an increase in the independent variable will be associated with an increase or a decrease in the dependent variable. That is, a negative regression coefficient indicates that an increase in the value of the independent variable will produce a decrease in the value of the dependent variable. Independent variable $X_6$ (average depth of exploratory wells) has an exponent of $-1.712$ which means that a 1 percent increase in the average depth of exploratory wells will produce a 1.7 percent decrease in $Y$ (the number of exploratory wells drilled).

Independent variable $X_7$ (time) is the only variable in equation 1 that appears as an exponent. The term, $0.896^{X_7}$ in the equation could be written $(1 - 0.1042)^{X_7}$; in this form, the term closely resembles the compound interest formula $(1 + i)^n$ with the exception of the algebraic sign. The negative sign makes this term assume the role of a compound discount formula. The meaning of $X_7$ (time) in this form in the equation is that when the other variables are held constant, the dependent variable $Y$ (the number of exploratory wells drilled each year) is reduced by 10.42 percent from the previous year. The Federal Power Commission staff had rationalized this treatment of "time" as follows:

One may speculate as to the reasons for this decline in drilling over time. One may point to the facts that since 1941 the number of allowable days for oil production in Texas has decreased, that the number of barrels of crude imported in the United States has greatly increased since 1940, that the amount of funds spent abroad by American petroleum companies for exploratory activity has increased .... We performed tests to see whether this constant rate of attrition in the number of exploratory wells over time shows any signs of slacking. Our tests indicate that there is no observable decrease in the rate of attrition.[3]

It should be noted here that the conceptual variable that $X_7$ represented was "other influences varying systematically with time"; and as such, it was forced to assume a broad catch-all type category. Since the Office of Economics was therefore not able to provide a clear explanation of the "reasonableness" of the behavior of $X_7$, this variable was selected primarily through statistical analysis.

The meaning of each of the individual components of equation 1 has now been considered. The equation's form, variables, and regression coefficients have each been explained in detail; the next task is to explain the steps that the Office of Economics took to evaluate the accuracy of this equation, which is the core of the exploratory activities submodel. This submodel was developed from an analysis of 22 years of natural gas industry time series data through the multiple regression technique explained in Chapter 3. The association that existed between the independent variables and the dependent variables in the multiple regression equation was measured by deriving multiple correlation coefficients. The multiple correlation coefficient for the exploratory activity submodel was 0.997.[4] A coefficient of multiple correlation must lie between $-1$ and 1. The closer it is to an absolute value of 1 the better the linear (here the log-linear) relationship between the variables, and the closer it is to 0 the poorer the linear relationship. The coefficient of multiple correlation squared is the ratio of "explained variation" to total variation, where "explained variation" refers to variations in the dependent variable associated with changes in the independent variables, and total variation refers to associated plus unassociated changes in the dependent variable. Therefore a coefficient of multiple correlation of 0.997 implies a coefficient of multiple determination of 0.994, which may be interpreted to indicate that 99.4 percent of a variation in the dependent variable can be accounted for by the multiple regression equation.

Another test of the accuracy of the exploratory activities submodel performed by the Office of Economics was a measure of the standard error of the estimate. This is a measure of the "scatter" of the actual values of the dependent variable. The standard error of the estimate for the exploratory model was computed to be 552 exploratory wells. This implies that approximately 68 percent of the sample points from a normally distributed sample should fall within plus or minus one standard error of the estimate of the estimated regression value. In effect, this measure of accuracy forms a zone around the regression estimates. Since the average number of exploratory wells drilled per year during the relevant period was in excess of 10,000, the model builders considered equation 1 to be a highly accurate replica of exploratory experience from 1940–1961.[5]

The statistical tests described above were employed to evaluate the accuracy of the entire regression equation as a unit without specific reference to the individual contribution of each independent variable. It should be noted that it is seldom difficult to develop a highly accurate multiple regression equation if enough independent variables are built into the estimating mechanism. The major usefulness of a multiple regression equation lies in its accuracy in determining the degree of influence that each independent variable has upon the dependent variable. The test employed by the Office of Economics to measure the accuracy of the regression coefficients was Student's $t$ test—a procedure that tests the significance of the partial correlation coefficients for each independent variable. The coefficient of partial correlation takes into account the explained and unexplained variations that arise both from the particular independent variable and from other sources. In order to determine whether or not the true value of the coefficient of each independent variable was significantly greater than zero, Student's $t$ test was performed for each. This test gives the probability of drawing a sample from a population that has a zero correlation, a data set that produces a coefficient as high as that actually obtained in the subject regression equation. If this calculated probability from the $t$ test is quite small, the coefficient is judged to be statistically significant; but if the probability is large there can be little confidence in the reliability of the coefficient.

The application of Student's $t$ test to the coefficients of the independent variables in the exploratory activities submodel provided the results shown in Table 4-2. These values indicate that the first variable—the initial base

**Table 4–2. Probability of Obtaining the Calculated Coefficients from a Zero Partial Correlation Population: Variables for Equation 1**

| Independent Variable | Probability |
|---|---|
| *Equation* 1: *Number of exploratory wells* | |
| $X_1$ (initial base gas price) | 0.08 |
| $X_2$ (wellhead price of crude) | 0.008 |
| $X_3$ (marketed gas production) | 0.0005 |
| $X_4$ (production of crude) | 0.0005 |
| $X_5$ (success ratio) | 0.0005 |
| $X_6$ (average depth of exploratory well) | 0.002 |
| $X_7$ (time) | 0.0005 |

*Source:* Harold H. Wein, *Natural Gas Supply and Demand* (Washington, D.C.: Federal Power Commission Docket No. AR 61–1, 1963), p. 49.

price of gas—exhibits the largest probability that its coefficient value could have been drawn from a population with a zero partial correlation. The chances are approximately 8 out of 100 that the coefficient value of 0.162 for $X_1$ could have been calculated if the true effect of the initial base price of gas on the dependent variable were actually zero. The calculated probabilities for the other independent variables were much lower, however; and more confidence was placed in their influence on the dependent variable. However, as will be explained later, the above should not be interpreted to mean that these probabilities support the exact values of the regression coefficients.

The Office of Economics interpreted these probabilities as easily supporting the coefficient values for six of the independent variables in equation 1. But in view of the relatively high probability that variable $X_1$'s measured coefficient was not its true value, two alternative hypotheses were tested. The first hypothesis stated that the true value of the regression coefficient was twice that which was initially observed; that is, that it was 0.324 rather than 0.162. The second hypothesis was that the regression coefficient of $X_1$ was 1 rather than 0.162. The chance of accepting a measured coefficient of 0.162

when the true value was 0.324 was calculated as 0.08, and the chance of accepting 0.162 with the second hypothesis was estimated to be 0.0005. From these tests it was concluded that:

It seems fairly clear that the observed elasticity (regression coefficient) of the initial base price of gas ($X_1$) with regards to the level of exploratory activity is very low, when other factors in the exploratory model are considered.[6]

Equation 1 was the essential core of the exploratory activities submodel. But this equation estimated only the number of exploratory wells drilled per year from given values of the independent variables discussed above, and estimates of drilling activity for hydrocarbons do not provide a sufficient basis for regulatory guidance. From the point of view of the model builders, those charged with the responsibility of setting prices should have additional knowledge concerning the factors that link base prices with new gas reserves. This additional information was provided by the other equations in the exploratory activities submodel. The final output from this submodel therefore included measures of annual additions to gas reserves and estimated total reserves. This sequence is illustrated in the flow chart where the estimated number of new exploratory wells as determined from equation 1 is converted by equation 8 into the number of exploratory feet drilled. Exploratory footage is in turn converted into gross additions to gas reserves by equation 9. Equation 10 accepts the gross additions to reserves as estimated by equation 9 and provides a final value for estimated total recoverable reserves.

It is significant that the exploratory activities submodel did not make specific provision for the hypothesis of directional exploration. The hypothesis of directional exploration holds that it is possible for producers to direct their exploratory drilling to either gas or oil and to know in advance with high accuracy whether a productive exploratory well will be oil or gas. This submodel therefore did not permit an evaluation of the range of possibilities which might occur if producers decided to narrow their exploratory activity by searching primarily for oil if regulated gas prices were thought to be too low for natural gas if high gas prices were set as an inducement. In the terminology of the chief of the Office of Economics the model was "neutral" with respect to the directional hypothesis.[7] The decision not to recognize the hypothesis of directional exploration and account for it in the exploratory activities submodel later proved to be the focal point of criticism concerning the validity and usefulness of the econometric model. This criticism will be considered later in this chapter.

The exploratory activities submodel is joined to the demand submodel by a linkage submodel. This linkage is accomplished by equation 2 which accepts the initial base price of gas (the input of equation 1) and converts this value into the average wellhead price of gas. The initial base price of gas employed in the exploratory activities submodel was the price of gas sales under new contracts, but the average wellhead price of gas must be derived from a mixture of many prices reflecting contract prices established in different time periods. Current price is required for the exploratory activities submodel because it is assumed that this is the price producers expect to receive from new production, but those who consume natural gas may find a large difference in new contract prices and current burner-tip prices. Equation 2 was developed by assuming that:

The Commission will not permit the prices of old contracts to be above new contracts at any of the five wellhead prices of new contracts we explored. Hence the average field price will always be below the new contract price over the period of projection.[8]

This means that a ceiling price set by the commission for new contracts lower than the current average field price would require a roll back in old gas prices. It was further assumed that approximately 6 percent of annual gas production was derived from new contracts in deriving the weighting factor in equation 2.

The demand submodels begin with equations 3 and 4 which convert the average wellhead price of gas into residential–commercial and industrial gas prices. Both of these equations indicate that increases in wellhead gas price are more than passed along to gas consumers. Equations 3 and 4 suggest that a 1 cent increase in the average wellhead price of gas would raise industrial gas price by 1.1 cents and raise residential–commercial by 1.5 cents. These equations also indicate that burner-tip prices are a positive function of the passage of time.

The core equations of the demand submodels—equations 5 and 6—are analogous in form to equation 2. These equations were derived through the same statistical technique used to develop equation 2 with the previously noted exception that the demand submodels were based on cross-section rather than on time-series data. The regression coefficients of equations 5 and 6 should be interpreted in the same fashion as the examples given for the exploratory activities submodel, and these equations were subjected to the same statistical tests. The multiple correlation coefficients for equations 5 and 6 were calculated as 0.989 and 0.918 respectively, and the results of Student's $t$ test are shown in Table 4–3. From these values it may be inferred that the significance of the regression coefficients of $X_{10}$ in equation 5 and $X_{12}$ in equation 6 were highly questionable. Both variables were retained in these equations, however, because they met the Office of Economics' conceptual test

**Table 4–3. Probability of Obtaining the Calculated Coefficients from a Zero Partial Correlation Population: Consumption Variables**

| Independent Variable | Probability |
|---|---|
| *Equation 5: Residential-Commercial Consumption* | |
| $X_9$ (residential–commercial gas price) | 0.00001 |
| $X_{10}$ (residential–commercial oil price) | 0.55 |
| $X_{11}$ (residential–commercial customers) | 0.0000001 |
| $X_{12}$ (degree days) | 0.001 |
| | |
| *Equation 6: Industrial Consumption* | |
| $X_{13}$ (industrial gas price) | 0.00001 |
| $X_{14}$ (industrial oil price) | 0.180 |
| $X_{15}$ (industrial coal price) | 0.015 |
| $X_{16}$ (industrial employment) | 0.005 |
| $X_{17}$ (steam electric generation) | 0.010 |
| $X_{12}$ (degree days) | 0.420 |

*Source:* Harold H. Wein, *Natural Gas Supply and Demand* (Washington, D.C.: Federal Power Commission, Docket No. AR 61–1, 1963), p. 100.

of reasonableness and because of the comparatively small values of their coefficients. The large negative regression coefficient values for the gas price variables in these two equations indicate that gas price increases to consumers tend significantly to reduce residential–commercial and industrial consumption, assuming no changes in the other variables.

The demand submodel's projections are completed by adding the two estimates derived from equations 5 and 6 to obtain estimated total national gas consumption. National gas consumption is then adjusted by equation 7 to produce marketed production of gas. The difference between gas consumption and marketed production lies in recognition that gas consumption is usually defined to exclude field use of natural gas for compressors, motors, and carbon-black plants, whereas marketed production includes these uses. Equation 7, therefore, increases national gas consumption estimates to arrive at a value for production. The output from equation 7 is then fed back into the exploratory activities submodel through equations 1 and 10. This feedback relationship closes the model. Now that all of the components of the model have been presented individually, the next task is to consider the staff's conclusions, which were drawn from the model as a unit and introduced into the *Permian* hearing.

The most important general finding derived from the model was that past increases in wellhead prices had not resulted in additional new gas reserves, and the model projections indicated (with all other relevant factors constant) that future price increases would lessen exploration and would reduce new additions to reserves in the long run.

This conclusion was reached by inputting a range of initial base gas prices into the model and comparing the results. This finding may be supported by comparing the magnitudes and signs of the regression coefficients or elasticities of the variables which represent natural gas prices in equations 2, 5, and 6. These equations indicate that exploration activity increases only slightly in response to price hikes while the consumption response is to lessen gas volume demanded significantly.

A second conclusion drawn by the staff from its experience with the model was that the commission cannot choose the "life index" of the country as a whole and the wellhead price simultaneously. Life index is a value which is derived by dividing the total reported reserves of gas in the United States at the end of a given year by the marketed production of gas during that year. The life index is thus a rough index of the number of years that the current reserves, without augmentation, could supply current requirements of gas. The life index (or the reserves to production ratio) is used as an inventory measure rather than as a method to attempt to predict a possible time when natural gas sources will be exhausted. Natural gas as a reserve is not in any serious danger of depletion.[10] The importance of the life index lies in its relationship to the lead time required by exploration teams to find productive new wells. The model demonstrates that wellhead price and life index are closely linked and are not independent; therefore, the selection of a value for one automatically determines the other. Another conclusion reached by the model developers was that increased sales of gas have been a result of the rapid growth of the economy combined with the displacement of other fuels. These growth forces have been so strong, it was argued, that they have outweighed the depressing effects that resulted from increasing gas prices. It was further asserted in the hearing that:

These growth factors, particularly on the demand side, will be far less vigorous in the future than they have been in the past, particularly since the level of industrial gas prices has now reached the point where competition from other fuels is keenly felt.[10]

From this observation it was suggested that the only way in which the gas industry could achieve a favorable growth rate in future years would be to reduce gas prices from their 1961 levels.

In summary, the position taken by those responsible for the model was that the commission should not act prior to a careful consideration of the effects that a range of prices might have on growth of the entire gas industry and on the long-run interests of the economy. Through the vehicle of their econometric model the staff predicted serious consequences following any decision that would permit an increase in wellhead price if the prices of other fuels remained relatively constant. The model was further used to support the contention that the commission could not independently determine both the wellhead price of gas and the value for the industry's life index because of fundamental interdependencies.

These conclusions were not acceptable to the respondents in this proceeding. Virtually every component of the econometric model was challenged by a spectrum of experts who could each claim far greater training and experience in a specific area than that possessed by Dr. Wein, who had introduced the model and was charged with its defense. The respondents supported a massive but scattered attack by experts who levied a variety of criticisms against the model. These criticisms will be considered in the following sequence of categories: false sophistication in model presentation; lack of professional performance on the part of Dr. Wein; faulty use of data; problems relating to equations and variables; statistical problems inherent in the model; and problems inherent in the fundamental concept of the model.

The respondents' charge that the staff's econometric model was initially presented in a tone of superficial sophistication was an important observation. The precision and confidence that characterized Dr. Wein's prepared statement did not remain intact during the hearing. The claim of complete objectivity in the model's development was seriously questioned and the respondents raised objections to the "specious air of precision" given to the regression coefficients by carrying their values in final form to three decimal places.

The criticisms leveled at the professional competence of Dr. Wein were based on two points; one, that Dr. Wein lacked specific ability as an econometrician, and two, that he lacked sufficient experience in the gas industry. Respondents answered the staff assertion that Dr. Wein had brought the new and powerful equipment of econometrics to bear on this problem by stating:

Econometric techniques are not new. In fact modern economics had advanced far beyond the least squares analysis used by Dr. Wein. . . . To demonstrate the invalidity of the Wein presentation, respondents presented highly qualified witnesses concerning all phases involved in the econometric study. Dr. A. P. Barton, who has spent much time constructing econometric models for the Central Planning Bureau of the Dutch government, demonstrated the invalidity of the statistical techniques employed by Dr. Wein. Certainly, a comparison of the qualifications of Dr. Barton with those of Dr. Wein demonstrates Dr. Barton's far greater expertise in this field, a fact clearly borne out by cross-examination of the two witnesses.[11]

The respondents' brief further indicated that Dr. Wein was not only inferior in ability in econometrics per se to Dr. Barton, the industry expert, but that he also knew less about computers than Mr. Willis, less about economic theory than Mr. Clark, less about economic geology than Mr. Allen, and less about oil and gas industry accounting than Mr. Wright. All of these cited experts testified for producers. It would appear that Dr. Wein was not able to establish himself as the leading expert in all fields relating to his testimony, but it should also be observed that respondents depended on a large number of experts all of whose expertise was claimed only in limited areas for attack on the Wein presentation.

The second general criticism against Dr. Wein was that he lacked experience in the gas producing industry. It was pointed out that Dr. Wein had joined the staff of the Federal Power Commission only one year prior to his testimony and that he had not had any prior experience in the oil and gas industry. The respondent's *Initial Brief* commented:

The construction of an econometric model is not a purely mechanical procedure, but in every stage involves economic decisions requiring judgment. The greater part of the information used is of a subjective type. Use of subjective judgment was necessary in the selection of the relationships in the various equations, in the selection of the explanatory variables, in the choice of the mathematical forms of the equations, and in the choice of particular data used. In each instance the choice can have a great effect upon the relationships which are being measured, and in most instances the choice required an intimate knowledge of the industry. Many of these choices which Dr. Wein actually made reveal his lack of knowledge of the industry.[12]

It was not denied that Dr. Wein had only brief experience in the oil and gas industry, and that the experience he did have was largely limited to his supervision of the development of the econometric model. During cross-examination it was apparent that Dr. Wein did not possess an exhaustive knowledge of oil and gas industry technology and terminology. However, to place this criticism in perspective, Dr. Wein did have available to him, while the model was under development, the expertise of the Federal Power Commission staff and consultants; therefore, the attack on the lack of experience of Dr. Wein does not imply that the model's developmental analysis was necessarily limited.

A third general category of criticism was that which centered on the data used in the construction of the model. Respondents not only suggested that the data employed were inaccurate; they also charged that data were arbitrarily selected to fit a preconceived bias on the part of the staff. Respondents pointed out that all the equations derived from time-series data did not have a common base. For example, equation 1 was based on 1940–1961; equation 1(a) was based on 1947–1961; equations 3 and 4 were based on 1946–1961; and equation 9 was based on 1947–1961 data with the omission of values for 1946–1954. Respondents suggested that these data bases reflected subjective judgment which was not sufficiently supported by reason to be acceptable. The respondents further demonstrated that by using data from different time periods the projections from the equations in the staff model could be made to vary widely, and that with differing data bases, the statistical tests described earlier would indicate a much larger likelihood of projection errors. Representatives of the gas industry did not agree with the reasoning of the staff as to why these different time periods were most representative for each of the equations. Criticism was also levied at equations 5 and 6 because they were based on cross-section rather than on time-series data which would have produced different results. Respondents further indicated that equation 5 used consumption data as reported by the Bureau of Mines, but measured number of customers from data supplied by the American Gas Association. It was pointed out that if both values had been taken from the same source, equation 5's coefficient values would clearly violate common sense and economic logic.

A fourth area for criticism by respondents centered on equation form and variable selection. It was pointed out that equations 1, 5, and 6 are multiplicative whereas equations 3 and 4 are linear. This particular point was carried no further, however, and the respondents did not indicate that they would have preferred all multiplicative or all linear forms. But criticisms against the procedures that determined the choice of variables, without doubt, had more substance than those against the form of the equations. It was developed through cross-examination that the variables in the model were essentially the only variables initially considered to be conceptually significant and were the only variables included in the original statistical analysis. Testimony indicated that these variables were picked in advance and that the model was built around them. Respondents stated that by ignoring such factors as the profitability of the oil and gas industry, alternative investment opportunities, the number of leasing prospects, the amount of unexplored acreage under lease, the imminence of lease expirations, the level of lease bonus and rental costs, and other such items, Dr. Wein had omitted consideration of some of the most fundamental factors on which industry decisions are based. Their *Initial Brief* stated with reference to this point: "bluntly speaking, Dr. Wein's imperious rejection of management judgment reflects on the credibility of his study."[13] Criticism was also leveled at the reasonableness of selected specific variables such as time $(X_7)$ and the success ratio $(X_5)$, which respondents asserted could not be supported through reason, logic, or experience.

Another major category of criticism against the model concerned its statistical shortcomings. Respondents' experts charged that Dr. Wein's equations suffered from unspecified degrees of autocorrelation, multi-collinearity, simultaneous equation bias, specification bias, and aggregation bias. A brief definition of these terms is in order. Autocorrelation is used to describe the lag correlation of a particular time series with itself, lagged by a number of time units. Autocorrelation is found in almost all data that displays trend or cyclical variation. Presence of autocorrelation may seriously influence tests of significance.

Multicollinearity is the tendency of many economic series to move together in the same pattern over time; it is the expression of common causes running through many economic variables. When a linear relation exists among two or more of the explanatory variables, it is difficult to measure their separate influence upon the explained variable.[14] This means that while a multiple regression equation is able to explain variations in the dependent variable, it may not be possible to estimate the separate influence of each independent variable. There was agreement during cross-examination that equations 1, 3, and 4 suffered from the problems of autocorrelation and multicollinearity, but the severities of these problems were not fully assessed.

Simultaneous equation bias, specification bias, and aggregation bias in this context all refer to problems revolving about the assumption of one way causation, which is built into the least squares approach. The problems of bias may be very significant when there are obvious interactions rather than a one-way causal effect from one side of an equation to the other. The respondent's econometric expert, Dr. Barton, commented on the exhibits which indicated high multiple correlation coefficients and high confidence levels by stating that:

Statistical tests are negative. In any empirical research you can never confirm your ideas in a definite way. You can only reject your preconceived ideas on the basis of evidence.[15]

All of the above categories of criticism were of a less important nature than the final category which included charges against the fundamental concepts of the model. The critical points reviewed above were of a type which might have been corrected or negotiated through agreements among the interested parties, but the following points were much more fundamental. The most damaging factor in the use of the staff model for this proceeding was its treatment of the directionality hypothesis. Dr. Wein stated that his model was neutral toward this hypothesis, but respondents' experts were able to establish that the Wein model implicitly assumed a type of negative directionality or a conscious refraining from exploration of gas when field prices rise. The econometric model's mathematical sequence indicates that an increase in the field price causes burner-tip prices to rise. This result, in turn, causes consumption to decrease and production to fall, and this drop in production will cause exploration to decrease to a greater extent than the higher prices cause exploration to increase. But if this decline in exploration is undifferentiated, then the increase in the field price of gas will ultimately force down the level of oil reserve additions. The *Joint Initial Brief* summed up the respondents argument on this point by stating: "to the extent that oil is the major product in relation to associated and dissolved gas, the tail is wagging the dog."[16] It was convincingly demonstrated during the hearing proceedings that the staff model was not neutral with respect to the directionality hypothesis, and Dr. Wein during cross-examination testified that an entirely new exploratory activities submodel would have to be developed if the directionality hypothesis were accepted. The significance of this admission is illustrated by reference to a portion of Chairman Joseph C. Swidler's review of the presiding examiner's opinion:

The touchstone of the presiding examiner's decision is his conclusion that the industry is increasingly becoming able to direct its exploration efforts toward finding new gas wells as distinguished from finding gas as a by-product in the search for oil. This newly documented ability of the industry to channel its exploration investment toward either gas or oil serves as the basis for a pricing system which we here adopt.

A second major criticism leveled at the concept of the model involved the problem of relating the national econometric model developed by the staff to the specific environment of the *Permian Basin* hearing. In the *Joint Reply Brief* the respondents summarized this criticism by stating that:

Dr. Wein on cross-examination was unable to tell us how his equations could be related to demand for gas from the Permian Basin or to Permian Basin supply. In fact, Dr. Wein stated that to really measure effects of price changes in the Permian Basin would have required 86 equations instead of the two which he actually used .... staff has now abandoned any pretense that that this study is related to Permian Basin prices.

The respondents were highly successful in arguing that this national model was out of place in this area rate proceeding—especially since the staff could not directly relate, through the model, the national to the area markets.

This chapter has described in analytical detail the composition of the staff econometric model. The findings that were developed from the applications of the model were presented, and the criticisms of the respondent parties were summarized and evaluated. The concluding exposition in this chapter indicated that the model was not judged to be an acceptable tool for helping the examiner reach his rate decision. The next step in this investigation is therefore an inquiry into the processes of model revision that the Office of Economics undertook in order to make the model more responsive to commission requirements. The following chapter evaluates the revisions of the model which were undertaken as a result of criticisms received and the experience gained in the *Permian* and subsequent area rate cases.

# Appendix to Chapter 4

The following equations comprise the econometric model which was developed by the Federal Power Commission's staff for use in the *Permian Basin Area Rate Hearing.*

*Equation 1: Exploration Model*

$$Y = 0.141 X_1^{0.162} \ X_2^{0.522} \ X_3^{1.566} \ X_4^{1.774} \ X_5^{0.996} \ X_6^{-1.712} \ 0.896^{X_7}$$

where  $X_1$ = initial base price of natural gas, adjusted by wholesale prices (¢/Mcf)

$X_2$ = wellhead price of crude petroluem, adjusted by wholesale prices ($/bbl.)

$X_3$ = marketed production of natural gas (billions of Mcf)

$X_4$ = production of crude petroleum (billions of bbls.)

$X_5$ = success ratio (percent productive exploratory wells of total exploratory wells)

$X_6$ = average depth of exploratory wells (thousands of feet)

$X_7$ = time (year 1940 = 1)

$Y$ = AAPG exploratory wells (thousands of wells)

Correlation coefficient = 0.997

*Equation 1(a): Average Depth Trend*

$$X_6 = 5.125 - 4.209 \ (0.879)t$$

where  $t$ = time (year 1940 = 1)

$X_6$ = average depth of exploratory wells (thousands of feet)

*Equation 2: Price Linkage Relation*

$$X_8 = a_t \ X_{1i} + (1 - a_t)X_{2it}$$

where  $X_8$ = average wellhead price of Natural gas (¢/Mcf)

$a_1 = 0.060$ ($t$ = year − 1961 = 1, ..., 9)       $a_5 = 0.305$

$a_2 = 0.120$                                         $a_6 = 0.370$

$a_3 = 0.180$                                         $a_7 = 0.435$

$a_4 = 0.240$                                         $a_8 = 0.500$

$a_9 = 0.565$

$X_{1i}$ = ceiling wellhead price (¢/Mcf) of gas under assumption $i$,

$X_{1,1} = 12.0$

$X_{1,2} = 14.5$

$X_{1,3} = 17.0$

$X_{1,4} = 19.5$

$X_{1,5} = 22.0$

$X_{2it}$ = the average wellhead price (¢/Mcf) of gas delivered in the year $t$ under contracts effected before 1962, given an effective ceiling $X_{1i}$ on prices under such contracts

$X_{2,1,t} = 11.8$

$X_{2,2,t} = 14.0$

For $i = 3, 4, 5$

$X_{2,i,1} = 15.225$       $X_{2,i,6} = 15.800$

$X_{2,i,2} = 15.350$       $X_{2,i,7} = 15.900$

$X_{2,i,3} = 15.475$       $X_{2,i,8} = 16.000$

$X_{2,i,4} = 15.600$       $X_{2,i,9} = 16.100$

$X_{2,i,5} = 15.700$

*Equation 3: Price Model—Residential and Commercial*

$$Y = 36.016 + 1.49 X_8 + 1.948 X_7$$

where   $X_8$ = average wellhead price of natural gas (¢/Mcf)

$X_7$ = time (year 1940 = 1)

$Y$ = gas price to residential and commercial consumer (¢/Mcf)

Correlation coefficient = 0.985

*Equation 4:   Price Model—Industrial*

$$Y = 6.372 + 1.097 X_8 + 0.431 X_7$$

where       $X^8$ = average wellhead price of natural gas (¢/Mcf)

$X^7$ = time (year 1940 = 1)

$Y$ = gas price to industrial consumer (¢/Mcf)

Correlation coefficient = 0.994

*Equation 5:   Residential and commercial consumption model*

$$Y = 0.549 X_9^{-0.795} X_{10}^{0.128} X_{11}^{0.979} X_{12}^{0.252}$$

where $X_9$ = gas price to residential and commercial consumer (¢/Mcf)

$X_{10}$ = oil price to residential and commercial consumer (¢/gal.)

$X_{11}$ = number of residential and commercial customers (thousands)

$X_{12}$ = degree days

$Y$ = residential and commercial consumption of natural gas

Correlation coefficient = 0.989

*Equation 6: Industrial Gas Consumption Model*

$$Y = 10,755 X_{13}^{-2.546}\ X_{14}^{0.737}\ X_{15}^{0.870}\ X_{16}^{0.503}\ X_{17}^{0.319}\ X_{12}^{-0.210}$$

where $X_{13}$ = gas price to industrial consumer (¢/Mcf)

$X_{14}$ = oil price to industrial consumer ($/bbl.)

$X_{15}$ = coal price to industrial consumer ($/ton)

$X_{16}$ = industrial employment (thousands of persons)

$X_{17}$ = steam electric plant power generation (billions of kwh)

$X_{12}$ = degree days

$Y$ = industrial consumption of natural gas (millions of Mcf)

Correlation coefficient = 0.918

*Equation 7: Production/Consumption Relation*

$$X^3 = R_{p/c}\, X_o$$

where $R_{p/c}$ = ratio marketed production of natural gas/consumption of natural gas

$X_o$ = consumption of natural gas (billions of Mcf)

$X^3$ = marketed production of natural gas (billions of Mcf)

*Equation 8: Exploratory Footage*

$$X_{19} = X_{18}\, X_6$$

where $X_{18}$ = number of exploratory wells (thousands of wells)

$X_6$ = average depth of exploratory wells (thousands of feet)

$X_{19}$ = exploratory footage (millions of feet)

*Equation 9: Additions to Reserves Model*

$$Y = 3.307 + 0.257X_{19}$$

where $\quad X_{19}$ = exploratory footage (millions of feet)
$Y$ = gross additions to reserves (billions of Mcf)
Correlation coefficient = 0.901

*Equation 10: Estimated Recoverable Reserves*

$$X_{21,t} = X_{21,t-1} + X_{20} - X_3$$

where $\quad X_{20}$ = gross additions to reserves of natural gas (billions of Mcf)
$X_3$ = marketed production of natural gas (billions of Mcf)
$X_{21,t}$ = estimated recoverable reserves of natural gas (billions of Mcf) in year $t$
$X_{21,1961}$ = 267.73 billions of Mcf
$X_{21,t-1}$ = reserves of gas in year $t-1$

# 5 Model Revision

The purpose of this chapter is to evaluate the efforts made by the Office of Economics to revise the *Permian* econometric model. This evaluation will be accomplished by examining the two econometric models used by the staff in the *Southern Louisiana Area Rate* case—the second Federal Power Commission proceeding involving an attempt to determine just and reasonable rates for natural gas on an area basis.

Testimony in the *Permian Area Rate Proceeding* was completed on January 7, 1964, and the *Presiding Examiner's Initial Decision* was published on September 17, 1964. The *Southern Louisiana* proceeding, which had been initiated in 1961, was in its closing stages at the time of completion of the *Permian* hearing. This second area rate case was undertaken concurrently with the first because, as the commissioners stated:

We do not believe that consistent with our statutory duties we can await determination of one area proceeding before we initiate others. ... Balancing the evils of delay against the possibility of perfection is always difficult.[1]

It is important to note that since the *Southern Louisiana* proceeding started before the *Permian* proceeding was completed, the Office of Economics did not have the opportunity to assess fully the effectiveness of the *Permian* econometric model in the first area rate case. The Office of Economics was aware, however, that those who had supported the model in the *Permian* hearing had fared poorly during cross-examination, as was indicated in the preceding chapter.

The econometric analysis that was placed in evidence by the staff in the *Southern Louisiana* proceeding was strictly within the broad philosophical framework of the so-called "Wein Model," the name given to the study of *Natural Gas Supply and Demand* presented by Dr. Harold H. Wein in the *Permian* proceeding. The *Southern Louisiana* economic exhibits were offered in support of the testimony by Mr. J. Harvey Edmonston, the Assistant Chief Economist and Chief Econometrician in the Office of Economics of the Federal Power Commission.[2] At the time of Mr. Edmonston's testimony, Dr. Wein was no longer associated with the Federal Power Commission. However, Mr. Edmonston, who was first employed by the Federal Power Commission in April 1962, had worked with Dr. Wein on the *Permian* model and had testified briefly during the *Permian* hearing concerning certain aspects of the directionality hypothesis and also with respect to the specific statistical tests used in the attempt to justify the Wein model.

The econometric presentation by Mr. Edmonston, in the *Southern Louisiana* proceeding included a national model and a *Southern Louisiana* model. Each model served a specific purpose. The national model was introduced because the staff argued, as it had in the *Permian* proceeding, that area regulation of ceiling prices on natural gas should be conducted with a frame of reference of national gas industry demand and supply factors. The *Southern Louisiana* econometric model was presented to provide a basis for making projections for the specific area involved in the hearing. It has been indicated in the preceding chapter that a major criticism of the *Permian* model was that its projections—which were made on a national basis—could not be directly related to the *Permian* geographical area. The new *Southern Louisiana* models, however, were intended to be used to make all the projections that would be required in the proceeding.

The national econometric model presented in the *Southern Louisiana* hearing was almost identical in structure to the *Permian* model. The national model, which when first introduced assumed a nondirectional capability in the search for hydrocarbons, was utilized to project national values for the following variables: exploratory activity of number of wells drilled, average depth of exploratory wells, total exploratory footage, natural gas reserves added, estimated recoverable reserves of natural gas, reserve–production ratios for natural gas, average wellhead field prices for natural gas, burner-tip prices for residential–commercial and industrial natural gas users, and the national production of natural gas. These output categories were the same as those of the *Permian* model; a minor difference was that the reserve–production ratio or life index was estimated as a specific variable in the *Southern Louisiana* national model while its estimation was not built into the formal structure of the *Permian* model. The *Southern Louisiana* national model retained the feedback mechanisms that related consumption and production to exploration which had been built into the parent *Permian* model. This feedback relationship " closed " the national model as explained in the preceding chapter.

The regional model was developed to relate national projections to the specific issue of the *Southern Louisiana* proceeding. This model, which will be referred to as the *Southern Louisiana* demand model, focused primarily on demand, and it did not include any feedback relationship. This procedure was in contrast to that used in the national model—which was designed to explain both supply and demand relationships. This approach was followed because those in the Office of Economics responsible for developing the econometric presentation for this proceeding had proceeded under the simplifying assumption that the intensity of exploratory activity could best be explained on the national level. Therefore, the model builders reasoned, if exploratory activity was chiefly influenced by national, rather than by regional factors, models developed for regional application should stress regional consumption variables and rely on the broad exploratory activity estimates derived from the national model.

The *Southern Louisiana* demand model was no more than a slightly modified version of the linkage submodel and the demand submodels of the *Southern Louisiana* national econometric model. The modification of the linkage submodel was of a definitional nature and the modification of the demand submodel involved only a simple weighting adjustment in the burner-tip price estimating equations. These adjustments were undertaken to give the *Southern Louisiana* demand submodel a capability for making consumption projections that would illustrate the effects of changing only South Louisiana area ceiling prices while holding all other gas prices constant at their 1962 level. These estimates were made on a state-by-state basis by relating the average burner-tip price in each state with hypothetical South Louisiana gas prices. The amount of average price change predicted for each state following a change in the South Louisiana ceiling depended on the proportion of South Louisiana area gas consumed. The *Southern Louisiana* demand model employed the same consumption equations as the *Southern Louisiana* national model. The method used to project area consumption involved several steps. The first was to predict national demand by states for given time periods; second, these values were then reduced by multiplying each state's total consumption by the fractional amount supplied to each state by the South Louisiana area in 1961, as compared to total gas consumption; and third, the total predicted consumption of South Louisiana gas was computed as equal to the sum of the Louisiana shares of state consumption values as determined in step two.

From the above description of the *Southern Louisiana* demand model, which was purportedly developed to relate the national with the regional markets, it should be obvious that only the most rudimentary methodology was employed. No new independent variables were included in the regional model, and the regional model, as developed, did not include any regional exploratory considerations.

Up to this point, this chapter has been concerned with giving a general introduction to the models employed in the *Southern Louisiana* case. Our investigation will now be directed to an analysis of the specific differences between the *Permian* and the *Southern Louisiana* models. This comparison requires the placing of the most important equations in each model in juxta-position and in order, to determine the significant points of difference. Equations will be coded so that equation number 1–P will be used to identify equation 1 of the *Permian* model, label 1–SL$_n$ refers to equation 1 of the *Southern Louisiana* model, and label 1–SL$_a$ refers to equation 1 of the *Southern Louisiana* area (demand) model.

*Equation 1–P*: $Y = 0.141 X_1^{0.162} \ X_2^{0.522} \ X_3^{1.566} \ X_4^{1.774} \ X_5^{0.996} \ X_6^{-1.712} \ 0.896^{X_7}$

*Equation 1–SL$_n$*:

$$Y = 0.135 X_1^{0.180} \ X_2^{0.515} \ X_3^{1.573} \ X_4^{1.818} \ X_5^{1.028} \ X_6^{-1.779} \ 0.894^{X_7}$$

where  $X_1$ = initial base price of natural gas, adjusted by wholesale price index (¢/Mcf)

$X_2$ = wellhead price of crude petroleum, adjusted by wholesale price index ($/bbl.)

$X_3$ = marketed production of natural gas (billions of Mcf)

$X_4$ = production of crude petroleum (billions of bbls.)

$X_5$ = success ratio (percentage of wells productive)

$X_6$ = average depth of exploratory wells (thousands of feet)

$X_7$ = time

$Y$ = exploratory wells (thousands of wells)

The values for the variables are the same for each equation with the exception of $X_6$. $X_6$ was determined as noted in Chapter 4 for the *Permian* model by equation 1(a) as:

$$X_6 = 5.125 - 4.209(0.879)^t$$

where

$$t = \text{time } (1940 = 1)$$

In the modified *Southern Louisiana* model this equation appeared as:

$$X_6 = 6.984(1.0023)^t - 2.524(0.9472)^t$$

where

$$t = \text{time } (1940 = 1)$$

The difference in these two subequations (which estimated average depth of exploratory wells) is simply that the *Permian* equation estimated that the average depth of exploratory wells would increase at an increasing rate with the passage of time after 1941, while the *Southern Louisiana* model estimated that average depth of exploratory wells would increase at a decreasing rate after 1947.

It may be observed that any changes in the regression coefficients of the independent variables in each of the equations used to estimate exploratory activity were insignificant. Therefore, the conclusions drawn from equation 1 in the *Permian* proceeding were reintroduced in total in the *Southern Louisiana* proceeding.

Moving in sequence through the exploratory activities submodel, in accordance with the logic previously illustrated in Figure 4–1, the next equation to be considered is equation 8. This equation estimated exploratory footage drilled. This definitional equation was not changed in the *Southern Louisiana* proceeding and is therefore exactly the same in that model as the one found in the appendix to Chapter 4. Equation 9, which estimated yearly additions to gas reserves as a linear function of exploratory footage, was changed only by increasing the constant to update the equation by one year. Equation 10, which estimated recoverable reserves, was identical in both models. The *Southern Louisiana* model had an equation 11, which was omitted from the *Permian* model; equation 11 was used to estimate the reserve production ratio or life index of natural gas. This ratio, which was discussed in Chapter 4, was derived by dividing the total estimated recoverable reserves of natural gas by the marketed production of natural gas for a given year. In completing the comparison between the exploratory activities submodels of the *Permian* and *Southern Louisiana* econometric models, it may be concluded that they were virtually identical.

Equation 2, which is the core equation of the linkage submodel, was identical in all models.

*Equation 2*:

$$X_8 = a_t X_{1i} + (1 - a_t)X_2$$

where $\quad t = 1\text{--}8$

$\quad\quad a_1 = 0.06$

$\quad\quad a_2 = 0.12$

$\quad\quad a_3 = 0.18$

$\quad\quad X_{1i}$ = candidate wellhead ceiling prices of gas

$\quad\quad X_{2it}$ = average wellhead price (¢/Mcf) of gas delivered in year $t$ under contracts enacted before 1962 for Permian and 1963 for Southern Louisiana—given that $X_{1i}$ is the effective ceiling on prices under such contracts.

$\quad\quad X_8$ = average wellhead price of natural gas

This equation (which was discussed in the preceding chapter) converts the current base price of natural gas into the average field price. The value of 0.06 for term $a_1$ indicates that in any given year about 6 percent of gas consumed was supplied from a contract negotiated during that year. The term $X_{1i}$, the ceiling wellhead price, was given a set of five different values during the process of testing the sensitivity of the models to changes in the price of gas. The term $X_{2it}$ is the average old contract price resulting from negotiations prior to the determination of the new ceiling price.

A major difference between the consumption submodels is found by comparing *Permian* equations 3 and 4 with the corresponding *Southern Louisiana* equations. In the *Permian* model, equations 3 and 4 converted the average wellhead price of gas as estimated by equation 2 into the residential–commercial burner-tip gas price and the industrial burner-tip gas price through the following:

*Euqation 3–P:*

$$Y = 36.106 + 1.49X_8 + 1.948X_7$$

where   $X_8$ = average wellhead price of gas (¢/Mcf)
   $X_7$ = time (1940 = 1)
   $Y$ = gas price to residential and commercial consumer

*Equation 4–P:*

$$Y = 6.372 + 1.097X_8 + 0.431X_7$$

where   $X_8$ = average wellhead price of gas (¢/Mcf)
   $X_7$ = time (1940 = 1)
   $Y$ = gas price to industrial customer

These two *Permian* equations, which were derived by a multiple regression analysis, suggest that an increase in $X_8$, the average wellhead price of gas, will be more than passed along to consumers. Equation 3–P is used to estimate that for every 1¢ increase in the wellhead price there will be a minimum increase of 1.5¢ in the burner-tip price to residential–commercial consumers and an increase of 1.1¢ in the burner-tip price to industrial customers. Dr. Wein, in his *Permian* testimony, had commented on this point as follows:

In an unregulated market this sort of differential is what one would expect in view of the inelasticity of residential–commercial demand compared to the high elasticity of industrial demand. A seller who has suffered additional costs would attempt to pass these to customers whose demand is not very responsive to price. According to our equations they have been successful in doing this.[3]

In contrast to equations 3 and 4 of the *Permian* model, equations 3 and 4 of the *Southern Louisiana* models were not derived as a result of statistical analysis.

*Equation 3–SL$_n$:*

$$Y_{nit} = Y_{1962} + X_{nit}$$

*Equation 3–SL$_a$:*

$$Y_{ait} = Y_{1962} + kX_{ait}$$

where $Y_{1962}$ = residential–commercial burner tip price (¢/Mcf) in 1962
$X_{nit}$ = estimated change in national average wellhead price of gas
$X_{ait}$ = estimated change in Southern Louisiana average wellhead price of gas
$Y_{nit}$ = national residential–commercial burner-tip price (¢/Mcf)
$Y_{ait}$ = South Louisiana residential–commercial burner-tip price (¢/Mcf)
$k$ = percent of natural gas from South Louisiana area consumed in a given state

Equations 4–SL$_n$ and 4–SL$_a$, which estimated the burner-tip prices for industrial consumers, were identical in form to equations 3–SL$_n$ and 3–SL$_a$.

*Equation 4–SL$_n$:*

$$Y_{nit} = Y_{1962} + X_{nit}$$

*Equation 4–SL$_a$:*

$$Y_{ait} = Y_{1962} = kX_{ait}$$

where $Y_{1962}$ = industrial burner-tip price 1962 (¢/Mcf)

$X_{nit}$ = estimated change in national average wellhead price of gas (¢/Mcf)

$X_{ait}$ = estimated change in South Louisiana wellhead price of gas (¢/Mcf)

$Y_{nit}$ = national industrial burner-tip price (¢/Mcf)

$Y_{ait}$ = South Louisiana industrial burner-tip price (¢/Mcf)

$k$ = percent of natural gas from South Louisiana area consumed in a given state

Equations 3 and 4 of the *Southern Louisiana* models simply assume a one-for-one relation between price changes in the average wellhead price for gas and changes in consumer burner-tip prices. That is, given the current burner-tip price, a 1¢ increase in the average wellhead price will cause a 1¢ increase in each of the two consumer burner-tip prices. This relationship is strikingly different from those derived by means of equations in 3 and 4 of the *Permian* model. The one-for-one relationships employed in the *Southern Louisiana* models implicitly assume nearly perfect regulatory control from the federal level through the local level. This difference in equations 3 and 4 between the *Southern Louisiana* model and in the *Permian* model is the most notable difference between the two models.

Since equations 5 and 6, which estimate residential–commercial and industrial consumption, for South Louisiana were not significantly different from their counterparts in the *Permian* model, there is no reason to examine them in detail. The independent variables used in equations 5 and 6 were used in both models, and only slight variations are found in the regression coefficients. These variations, which are not significant, were mainly due to the use of cross-section data for the year 1962 in the *Southern Louisiana* model while the *Permian* consumption equations were developed from 1961 data. Differences in the correlation coefficients for the independent variable "degree days" are explained by noting that the definition of this variable was changed in the *Southern Louisiana* model to make specific allowances for the proportion of gas used in space heating as opposed to the definition of this variable in the *Permian* model, which did not differentiate between weather effects on gas used for space heating and gas used for other purposes. This paragraph concludes the structural comparison of model equations used in the *Permian* and *Southern Louisiana* cases; however, for the convenience of the reader, all of the *Southern Louisiana* equations have been reproduced in the appendix to this chapter so that point-by-point comparisons can be made with the *Permian* equations in the appendix to Chapter 4. The following discussion provides a comparative analysis of the projections made by the

Office of Economics and introduced in support of testimony in the *Southern Louisiana* hearing.

Projections of gas consumption computed by using the *Southern Louisiana* equations of the national econometric model required that certain assumptions be made concerning the price behavior of competing fuels. First, a "proportional competitive price assumption" was made; that is, it was assumed that the prices of competing fuels moved proportionally with burner-tip prices in the residential–commercial and the industrial markets. Second, an alternative assumption, "the constant competitive price assumption," was made. In this case, it was assumed that the price of competing fuels remained constant at their 1962 levels.

Tables 5–1 and 5–2 illustrate the projections made for five different ceiling prices with each of these competitive fuel price assumptions. Table

**Table 5–1. Residential–Commercial and Industrial Consumption of Natural Gas as Projected by the *Southern Louisiana* Model Assuming Competing Fuel Prices Vary Proportionately with Gas (Billion Mcf)**

| Assumed National Average Ceiling Price (¢/Mcf) | 1963 | 1966 | 1970 |
|---|---|---|---|
| *Total Consumption* | | | |
| 10¢ | 15.58 | 17.84 | 20.47 |
| 12¢ | 14.11 | 16.10 | 18.42 |
| 17¢ | 12.10 | 13.49 | 15.10 |
| 22¢ | 11.96 | 12.92 | 13.86 |
| 24¢ | 11.91 | 12.71 | 13.43 |
| *Residential–Commercial Consumption* | | | |
| 10¢ | 4.80 | 5.19 | 5.72 |
| 12¢ | 4.72 | 5.11 | 5.63 |
| 17¢ | 4.59 | 4.95 | 5.43 |
| 22¢ | 4.58 | 4.90 | 5.33 |
| 24¢ | 4.57 | 4.89 | 5.29 |
| *Industrial Consumption* | | | |
| 10¢ | 10.78 | 12.65 | 14.75 |
| 12¢ | 9.39 | 10.99 | 12.79 |
| 17¢ | 7.51 | 8.54 | 9.68 |
| 22¢ | 7.39 | 8.02 | 8.53 |
| 24¢ | 7.34 | 7.82 | 8.15 |

*Source:* Federal Power Commission, *Southern Louisiana Area Exhibit* 38, Docket No. AR 61–2, 1964, pp. 15, 21, 27.

**Table 5–2. Total Consumption of Natural Gas as Projected by the *Southern Louisiana* Model Assuming Competing Fuel Prices Remaining at 1962 Levels (Billion Mcf)**

| Assumed National Average Ceiling Price (¢/Mcf) | 1963 | 1966 | 1970 |
|---|---|---|---|
| *Total Consumption* | | | |
| 10¢ | 20.25 | 23.31 | 26.82 |
| 12¢ | 16.44 | 18.79 | 21.50 |
| 17¢ | 12.12 | 13.26 | 14.59 |
| 22¢ | 11.87 | 12.25 | 12.50 |
| 24¢ | 11.77 | 11.89 | 11.83 |
| *Residential–Commercial Consumption* | | | |
| 10¢ | 4.86 | 5.77 | 5.80 |
| 12¢ | 4.76 | 5.16 | 5.68 |
| 17¢ | 4.59 | 4.94 | 5.41 |
| 22¢ | 4.58 | 4.88 | 5.28 |
| 24¢ | 4.57 | 4.86 | 5.23 |
| *Industrial Consumption* | | | |
| 10¢ | 15.38 | 18.04 | 21.02 |
| 12¢ | 11.67 | 13.63 | 15.82 |
| 17¢ | 7.53 | 8.32 | 9.17 |
| 22¢ | 7.29 | 7.37 | 7.21 |
| 24¢ | 7.20 | 7.03 | 6.60 |

*Source:* Federal Power Commission, *Southern Louisiana Area Exhibit* 38, Docket No. AR 61–2, 1964, pp. 15, 21.

5–1, which gives projections under "the proportional competitive price assumption," indicates that as national average ceiling prices were varied from 10 to 24¢ per Mcf, total projected consumption of natural gas ranged from 15.58 to 11.91 billion Mcf. For 1963, the projected drop in residential–commercial consumption was approximately 5 percent as price rose from 10 to 24¢ per Mcf, but the corresponding drop in industrial demand was about 31 percent. Table 5–2, which details projected consumption values under the "constant competitive price assumption," indicates a marked response by industry in shifting to, or away from, gas as its price varied while competing fuels were held at their 1962 levels. Residential–commercial reactions to price changes are slight, and in comparing this category with its counterpart in Table 5–1, one finds very little difference. With the " constant competitive price assumption," projected residential–commercial consumption dropped by 6 percent in 1963 as ceiling prices rose from 10 to 24¢, whereas industrial consumption dropped by more than 50 percent in 1963 and by almost 70 percent in 1970 with the same ceiling price increase. The model, therefore, indicates that the volume of industrial consumption of natural gas is highly dependent on the relative prices of competing fuels.

It is interesting to compare these projections of gas consumption with projections made on the basis of the *Permian* model. Table 5–3 illustrates

### Table 5–3. Total Residential–Commercial and Industrial Consumption of Natural Gas as Projected by the *Permian* Model Assuming Competitive Fuel Prices Vary Proportionately with Gas (Billion Mcf)

| Assumed National Average Ceiling Price (¢/Mcf) | 1963 | 1966 | 1970 |
|---|---|---|---|
| *Total Consumption* | | | |
| 12¢ | 13.74 | 15.54 | 17.81 |
| 17¢ | 12.14 | 13.51 | 15.26 |
| 22¢ | 11.93 | 12.93 | 14.14 |
| *Residential–Commercial Consumption* | | | |
| 12¢ | 4.75 | 5.25 | 5.92 |
| 17¢ | 4.55 | 5.01 | 5.62 |
| 22¢ | 4.52 | 4.92 | 5.45 |
| *Industrial Consumption* | | | |
| 12¢ | 8.99 | 10.29 | 11.89 |
| 17¢ | 7.59 | 8.50 | 9.64 |
| 22¢ | 7.41 | 8.01 | 8.69 |

*Source:* Federal Power Commission, *Permian Basin Area Exhibit* 235–C, Docket No. AR 61–1, 1963, pp. 30, 35, 40.

consumption estimates—assuming that competitive fuel prices vary proportionally—for natural gas, corresponding to selected ceiling prices for the years 1963, 1966, and 1970. The values in Table 5–3 correspond closely to those of Table 5–1. National estimates made by the use of each model for a given ceiling price are almost identical for residential–commercial consumption projections. The primary explanation for the very slight variations, more than likely, lies in multicollinearity rather than a similarity between equations 3 and 4 in the two models. These equations, which were discussed in detail above, convert the average wellhead price of gas into burner-tip prices for the two demand submodels. These equations in the *Permian* model were used to convert a 1¢ increase in the average wellhead price of gas into an approximate 1.5¢ increase in residential–commercial burner-tip prices and an approximate 1.1¢ increase in the industrial price. Equations 3 and 4 in the *Southern Louisiana* model, however, simply passed on wellhead price changes on a one-for-one basis. This meant that for the same given change in the average wellhead price of gas, the *Permian* model would estimate the volume of gas demanded with respect to higher consumer prices than would the *Southern Louisiana* model. It is not necessary to present a table of *Permian* model projections which corresponds to *Southern Louisiana* estimates made under "the constant competitive price assumption" because the same relationships exist under that assumption in the two models as exist under the "proportional price assumption."

By moving from the strictly national models to the *Southern Louisiana* demand model, projections were made to estimate the effects of each of a range of strictly area prices on national gas consumption. These projections were made by assuming that all other area gas prices remained constant at their 1962 levels while the South Louisiana area gas prices were permitted to vary from 13.7 to 27.7¢ per Mcf. The actual national average wellhead price in 1962 was approximately 17¢ per Mcf. Under "the proportional competitive price assumption," as the South Louisiana ceiling price rose from 13.7 to 27.7¢ per Mcf, the projected total national consumption of natural gas for 1963 fell from 12.54 to 12.05 billion Mcf, and the corresponding drop for 1970 was from 16.32 to 15.41 billion Mcf. With respect to total residential–commercial consumption, there was a 1 percent decline in 1963 and a 2 percent decline in 1970 in the amounts of gas which would be used in the national market when the South Louisiana ceiling price was changed from 13.7 to 27.7¢ per Mcf. The total industrial consumption estimates indicated corresponding declines of 5 percent for 1963 and 7 percent for 1970 for the same area price increase. This *Southern Louisiana* area demand model was also used to project the effects that various hypothetical area ceiling prices would have on the consumption of South Louisiana gas; however, it is not necessary to examine these estimates in detail because, as it will be demonstrated later, they were found to be not applicable to this rate proceeding.

The conclusions drawn by the staff on the basis of the econometric presentation in the *Southern Louisiana* proceeding was basically the same as those drawn from the use of the *Permian* model. These conclusions were extensions of the argument developed from the models that the granting of any increase in the initial base price of gas would seriously reduce the rate of increase in consumption of natural gas and would absolutely reduce future exploratory activity so that the number of exploratory wells drilled would decline and additions to reserves would fall.

The arguments denying these conclusions by those who represented the interests of the natural gas producers in the *Southern Louisiana* proceeding were essentially the same as those presented to discredit the *Permian* econometric analysis. Criticisms leveled at the staff's econometric presentation by the advocates representing the gas industry were based on the fundamental contentions that those analysts within the Federal Power Commission who had developed the models were professionally naive and were grossly lacking in experience in the oil and gas industry. The specific arguments developed were almost identical to those detailed in Chapter 4 and are therefore given only a brief review below.

The principal expert in econometrics for the gas interests was Dr. Cootner, an Associate Professor of Finance at the Massachusetts Institute of Technology. Dr. Cootner, who had been employed by the Amerada Petroleum Corporation to evaluate the staff's econometric findings, criticized the staff's models through an approach similar to that of Dr. Barton, who had acted as the industry's primary econometric expert in the *Permian* hearing. Dr. Cootner indicated that those who developed the econometric models presented by the staff had neglected consideration of many important managerial decision variables that might influence ultimate profitability. Dr. Cootner pointed out:

Any economic study of any private enterprise industry which concludes that investment will take place regardless of profitability is simply not worthy of serious consideration.[4]

When Mr. Edmonston was questioned as to why the staff models did not give consideration to levels of industrial profitability he replied:

Well, my previous experience suggested to me that it was not possible to attempt a real measure of profitability in the analysis. Maybe it should be there or if I had an adequate surrogate, then I did not know how to use it effectively.[5]

Dr. Cootner pointed out that since the staff model did not include variables to measure profitability, the models would indicate an ever-increasing exploratory activity as the ceiling price for gas was lowered. For example, he calculated that with a 2¢ per Mcf ceiling price the staff model would predict that in 1970 the industry would drill 268,310 exploratory wells. This figure was greater than the total of all exploratory wells drilled since 1930.[6]

The *Southern Louisiana* models were further criticized for not incorporating dynamic considerations in their demand submodels. And the projections of the demand submodels were challenged by another Amerada-sponsored expert—Mr. Sherman Clark of the Stanford Research Institute—who presented an analysis of the price elasticity of demand for gas. Mr. Clark asserted that the actual demand for gas was much less elastic over the relevant price range than the projections of the staff indicated. Mr. Clark indicated that the demand submodels were inaccurate because the estimating equations did not take sufficient account of the competition of electricity in the resident-ial–commercial market, and further that it was a serious error to use the same demand forecasting equation in all states when in fact the price elasticities of demand varied widely among the states.[7]

From the point of view of the hearing examiner, perhaps the most dam-aging attack on the staff's models during the proceeding was delivered by Dr. Cootner who set forth nine separate instances in which opposite results (from those reached by the staff) were obtained by using staff data with different economic and econometric approaches. These different approaches included using nonlinear estimation techniques rather than linear regression; using time series rather than cross-section analysis; combining cross-section and time series analysis; dividing of the states into groups and computing separate gas demand equations for each; and other similar technical variations. Following the development of this attack, industry advocates asserted:

. . . the results of the numerous alternative approaches and/or estimates used by Dr. Cootner lead to such uniformly opposite results than those presented by staff that one can reasonably inquire whether the formulation presented by staff is not the *only* formulation that will support their hypothesis; every other alternative approach and/or estimate leads to opposite results.[8]

It must be noted here that Dr. Cootner did not develop a complete model to replace the staff model; nor did he defend any of his nine alternatives to specific subcomponents of the staff's models. His attack was based solely on a strategy of discrediting the staff's analysis.

The basic fault in the *Southern Louisiana* econometric presentation was its use of an assumption regarding directional exploration. The *Southern Louisiana* and the *Permian* models had both rejected this hypothesis—which states that it is possible for producers to direct their exploratory drilling to either gas or oil. The *Permian* and *Southern Louisiana* models assumed instead that "exploratory effort is regarded as a search for hydrocarbons in general rather than for natural gas, petroleum, or natural gas liquids specifically."[9]

Shortly after the above testimony was given, *Federal Power Commission Opinion 468* was issued. It stated, in part, that since the industry did now have directional capability, the Wein studies could not be relied upon in the *Permian* case. Following this decision the hearing examiner in the *Southern Louisiana* proceeding asked the staff if it would be possible to revise quickly the econo-metric analysis and resubmit a model that assumed directionality. In response to this request, staff witness Edmonston presented an alternative econometric study and he testified with respect to this new study as follows:

Whereas, in my previous testimony, exploratory effort was regarded as a search for hydrocarbons in general . . . in the present analysis it is assumed that exploration for natural gas and oil may respond differently to economic factors, reflecting a geological capability and an economic motive to drill directionally. . . . When production costs are given, the rate of exploration for oil and gas is closely dependent on selling prices paid and the amounts produced and sold; and as selling prices to consumers rise, the amounts consumed, and hence produced, tend to decline. In contrast to the intuitively appealing view that higher prices will induce more exploratory activity, this hypothesis states that producers will not continue to explore more, even if prices are rising, if they cannot sell more.[10]

There was no change in the basic econometric arguments following the introduction of the new models. The staff contended that the general conclusions it had drawn from its original study could likewise be drawn from the new studies.

It is easy to understand why the conclusions drawn from the new model were the same as those obtained from the original model—which did not assume the directional hypothesis—when one examines the limited extent of the changes made in the revised model. The new model was changed from the one it replaced only through a redefinition of equation 1. In the original model, a single equation was used to estimate the total number of exploratory wells drilled; in the new model, this equation 1 was replaced by 1(g) and 1(o) to make estimates of the number of exploratory oil wells drilled.

*Equation 1(g):*

$$Y = 0.016 X_1^{0.034} \ X_2^{0.427} \ X_3^{0.929} \ X_{4a}^{0.538} \ X_5^{1.095} \ X_6^{-0.071} \ 0.950^{X_7}$$

All of the variables are the same in 1(g) as explained in the case of equation 1 earlier in this chapter, with the exception of $X_{4a}$. In equation 1, $X_4$ represented the production of crude petroleum in thousands of barrels; in equation 1(g) $X_{4a}$ represents the number of exploratory oil wells drilled in thousands of wells.

*Equation 1(o):*

$$Y = 0.059 X_2^{1.043} \ X_4^{4.691} \ 0.902^{X_7}$$

where
$X_2$ = wellhead price of crude petroleum
$X_4$ = production of crude petroleum
$X_7$ = time (current year—1939)
$Y$ = number of exploratory oil wells drilled

These new equations (which replaced equation 1) resulted in projections to the effect that changes in the price of gas would have even less effect on the number of gas wells drilled than originally estimated.

Econometric experts who were testifying for Amerada Petroleum were quick in seeking to discredit this new model which took account of directionality. It was pointed out that Mr. Edmonston's newly developed equations 1(g) and 1(o) had been developed by assuming perfect directionality for the entire 24-year period from which his equations were derived. The commission finding regarding directionality in the *Permian* proceeding, however, had been to the effect that directional capability in exploratory activities was a recent phenomenon. Advocates of the gas industry asserted that since the directionality hypothesis was of only recent applicability, the price elasticity of the search for gas was understated by equation 1(g). Amerada's experts further criticized the use of the two new equations by noting that the equation used to predict the number of exploratory gas wells had four more independent variables than did the parallel original equation used to predict the number of exploratory oil wells. When Mr. Edmonston was questioned about his choice of variables for equations 1(g) and 1(o), he was asked to explain why the predicted number of gas wells depended in part on the number of oil wells drilled while the number of oil wells drilled was not in any way related within the model's structure to exploratory gas activity. He responded:

My intuitive hunch was that the effect of gas search on oil discovery would be less than the reverse effect. As I have said before, however, I consider 1(o) by no means exhaustive. I didn't give it as critical attention—I didn't have time to—as I did 1(g).[11]

The influence on the ultimate outcome of the staff's econometric presentation in the *Southern Louisiana* proceeding was not much more significant than it had been in the *Permian*. The specific attacks on the econometric models never were referred to, dealt with, or answered in the staff briefs.[12] The staff's response to Dr. Cootner's criticisms (which had been characterized as "Amerada's devastating attack" by presiding examiner Zwerdling) was to request that the econometric experts, who were testifying for the gas interests, develop and present constructive alternatives to the staff models. The staff charged that Dr. Cootner's testimony did not imply that the staff econometric findings were invalid; rather, the staff argued, Cootner had only hypothesized that if the models had been developed in some other manner the resulting conclusions might have been different. The staff suggested that if he was interested in making a positive contribution to the proceedings, Dr. Cootner should develop his own model using his best professional judgment and experience.[13] However, the presiding examiner did not feel that an alternative model was necessary and dismissed the staff's request for such a model by stating:

This short-hand attempt by staff to dispose of and ignore the detailed and specific criticisms leveled at its ecometric studies is unimpressive and is not entitled to serious consideration . . . witness Cootner effectively demonstrated numerous flaws and inadequacies in the staff econometric models to such an extent as to require the conclusion that it would not be wise to place reliance on the staff's models in the instant proceeding. . . . The staff's contention that it was necessary for Dr. Cootner to develop an econometric model of his own, to support his criticisms, is entirely unpersuasive.[14]

Even though the examiner ruled that the staff's econometric presentation was not sufficiently reliable to play a persuasive role in this proceeding, it is most important to note that he did not assert that there was no useful role for econometric models in future rate proceedings. The examiner concluded his assessment of the econometric presentation by stating:

It is regarded as commendable that the Staff has devoted such vast time and effort in an attempt to develop its econometric studies as a useful tool to be employed in area rate proceedings. Staff witness Edmonston made a gallant effort in this direction. He courageously and conscientiously undertook and carried out, to the best of his ability, an immensely difficult and complex task, operating with limited resources and limited time. His responses during grueling and extensive cross-examination were marked by complete forthrightness and integrity. . . . It may properly be suggested that enough has already been demonstrated to indicate that econometric studies ultimately, on the basis of still further study and improvement by the Staff, may one day be brought to the point where they will provide a significant, useful and reliable tool in this connection—and it might therefore be worthwhile for the Staff to continue its efforts in this direction.[15]

Even though the econometric models did not directly influence the decision in the *Southern Louisiana* hearing, the econometric presentation did succeed in suggesting to the examiner that, with more refinement, econometric models could prove to be valuable tools in future regulatory proceedings. The encouragement given to the staff to continue its econometric investigations indicates an interest in this approach by the presiding examiner in the *Southern Louisiana* case which was not evidenced in the *Permian* case. The econometric presentation in the *Permian* proceeding was briefly mentioned in the *Examiner's Initial Decision* only for the purpose of dismissing it as being " not relevant or material to the problem."[16] In contrast to this cursory treatment in the *Permian Area Initial Decision*, 38 pages of summary and analysis in examiner Zwerdling's published decision were devoted to the *Southern Louisiana* econometric presentation. The preponderance of comments on the material presented in these 38 pages was not favorable to the staff's performance in the development and defense of the models; but, for the first time in a hearing examiner's initial decision, some degree of support and encouragement was given to the econometric approach. The staff, therefore, was encouraged to develop a new and much improved set of models for use in future regulatory proceedings. The Office of Economics is responsible for the development of such econometric presentations, and it has been devoting a respectable portion of its available analytical effort toward developing a post Wein–Edmonston econometric model which will hopefully prove to be a more reliable tool for regulatory application. Some of the more important points which should be considered in performing this current task are discussed below.

The *Southern Louisiana* model builders attempted to repair two of the basic conceptual faults that had been incorporated into the structure of the *Permian* model. The *Southern Louisiana* econometric presentation first attempted to relate the national gas market to the area market, and later an attempt was made to build the directionality hypothesis into the exploratory activities submodel. Neither of these patch up efforts was successful, however, for the reasons cited in the preceding discussion. It would appear that the Federal Power Commission model builders need to do more than repair; they need to reformulate their basic methodology. For example, future staff econometric models might rely less heavily on the statistical technique of multiple regression. This technique has been employed exclusively in the staff models to derive empirically based relationships. Some interesting comments on the use of multiple correlation analysis for projection purposes were made by Robert Goodell Brown:

I happen to think that, except in certain very special cases, the use of multiple regression models can be very dangerous. There appears to be no statistical test that will tell whether the model is a good one or not. . . . Even if two series are related, one doesn't know the independent series long enough in advance to be useful in a forecast. If one has to predict the independent series, any errors in that prediction will be amplified by the coefficients in the model. If the coefficients aren't large enough to amplify, then the terms aren't significant in the model and can be dropped.[17]

When the above caveat is linked to the quotation below, the reliability of forecasting techniques derived from multiple regression analyses of historically observed data is subject to increasing doubt. Mr. Brown continued:

The whole point of making a forecast, of course, is that one doesn't know what the future observations will be. The estimated values of coefficients are based on past data alone. They are the best possible values to use if one were to forecast backward in time. Let us look at some of the reasons why they may not be the best values to use in the future. . . .

Consider a physical process, like tracking the position of an aircraft. There is a maximum acceleration . . . a limited rate of climb . . . a limited rate of descent . . . a minimum radius for the tightest turn. If the aircraft has been maneuvering during the most recent observations . . . the coefficents in the model will not be an exact representation of the path of the aircraft. Therefore it is possible to compute . . . an absurd forecast of future positions.[18]

It should not be too hard to establish the premise that historical data relating to the parameters of the natural gas model describe an industry experiencing periods of significant change. The changes that have influenced gas production and marketing have been the consequence of such diverse factors as the rapid growth of pipeline sales following the second world war, the growth of federal regulatory activity in the natural gas area, the rise in importation of foreign crude oil, the recent capability of directional exploration, and other related phenomena. The gas production industry has apparently been subjected to a series of conflicting pressures affecting both its growth and contraction during the historical period for which the Federal Power Commission's Office of Economics has attempted to develop its projecting mechanism.

The suggestion that model builders may not have a suitable time series data base for model building should not be interpreted to mean that econometric models should not be used. This does suggest, however, that the model builders may have to rely less on pure statistical analysis and more on reason and experience. The introduction of a greater reliance on reason and experience into the development of a new econometric model would strip staff models of some of their previous sophistication and precision. However, these qualities which in cross-examination, were labeled "false sophistication" and "specious precision" were the prime targets of the expert witnesses for the gas producers. In the *Permian* and *Southern Louisiana* proceedings the successful attack on these points was instrumental in discrediting the econometric presentations. If the future econometric arguments developed by the staff are more general, then industry experts will feel greater pressures to develop their own models, which will produce forecasts more in line with producer interests. This strategy would then accomplish one of the goals sought by the staff in the *Southern Louisiana* proceeding—an evaluation of the merits of the staff econometric conclusions achieved by means of confrontation of econometric models.

In its work on the set of new models designed to replace the Wein–Edmonston analysis, the Office of Economics would have strengthened its presentation by placing greater emphasis on exploration and supply. Consideration in future work ought to be given to replacing the original dependent variable in the exploratory activities submodel found in the Wein–Edmonston models—exploratory wells drilled—with the dependent variable, additions to new reserves. This substitution in format would help bring analytical focus to bear on the desired end product of exploration—the addition of new gas reserves—and not the drilling of holes. This reformation of the supply sub-model should also give greater emphasis to the variables which bear on decisions to drill strictly wildcat gas wells, since these wells lead to the discovery of new fields. The Federal Power Commission staff must also engage expert consultants to assist them during model development and subsequent direct and cross-examination of its econometrics-related testimony in future proceedings.

This chapter has analyzed in detail the econometric models presented by the staff in the *Southern Louisiana* proceeding. The *Southern Louisiana* models were compared to the *Permian* model, and the degree of influence the *Southern Louisiana* models provided toward contributing to a more effective regulatory decision was evaluated. The materials presented in this chapter indicate that even though the econometric models did not contribute significantly to the *Southern Louisiana* proceeding the potential contribution of this approach was recognized by the presiding examiner. The chapter that follows builds on this potential. That is, chapter 6 explores the role that a more acceptable econometric model might assume in future regulatory proceedings and contains an evaluation of alternative strategies for staff's econometric presentations.

# Appendix to Chapter 5

## Glossary of Southern Louisiana Equations

The following equations comprise the econometric models which were developed by the Federal Power Commission's staff for use in the *Southern Louisiana Rate* hearing. The equations which follow are those which include the hypothesis of directionality.

*Equation 1(g): Exploratory Gas Wells*

$$Y = 0.016 X_1^{0.034} \ X_2^{0.427} \ X_3^{0.929} \ X_{4a}^{0.538} \ X_5^{-0.071} \ X_6^{1.095} \ 0.950^{X_7}$$

where $X_1$ = initial base price of natural gas, adjusted by wholesale prices (¢/Mcf)

$X_2$ = wellhead price of crude petroleum, adjusted by wholesale prices ($/bbl.)

$X_3$ = marketed production of natural gas (billions of Mcf)

$X_{4a}$ = AAPG exploratory oil wells (thousands of wells)

$X_5$ = exploratory gas well success ratio (percent productive exploratory gas wells of total exploratory wells)

$X_6$ = average depth of exploratory wells (thousands of feet)

$X_7$ = time (year 1940 = 1)

$Y$ = AAPG exploratory gas wells (thousands of wells)

Correlation coefficient = 0.999
Number of observation years = 24

*Equation 1(g)a: Average Depth Trend of Exploratory Gas Wells*

$$X_6 = 6.984(1.0023)^t - 2.524(0.9472)^t$$

where $t$ = time (year 1940 = 1)

$X_6$ = average depth of exploratory gas wells (thousands of feet)

*Equation 2: National Average Wellhead Price Relation*

$$Y = a_t X_{1i} + (1 - a_t) X_{2it}$$

63

where $Y$ = average wellhead price of natural gas (¢/Mcf)

| | |
|---|---|
| $a_1 = 0.060$ ($t$ = year − 1962 = 1, ..., 8) | $a_5 = 0.305$ |
| $a_2 = 0.120$ | $a_6 = 0.370$ |
| $a_3 = 0.180$ | $a_7 = 0.435$ |
| $a_4 = 0.240$ | $a_8 = 0.500$ |

$X_{1i}$ = ceiling wellhead price of natural gas (¢/Mcf) under assumption $i$,

| | |
|---|---|
| $X_{1,1} = 10$ | $X_{1,4} = 22$ |
| $X_{1,2} = 12$ | $X_{1,5} = 24$ |
| $X_{1,3} = 17$ | |

$X_{2it}$ = the average wellhead price of gas (¢/Mcf) delivered in the year $t$ under contracts effected before 1963, given an effective ceiling $X_{1i}$ on prices under such contracts

$X_{2,1,t} = 9.9$

$X_{2,2,t} = 11.8$

For $i$ = 3, 4, 5

| | |
|---|---|
| $X_{2,i,1} = 15.350$ | $X_{2,i,5} = 15.800$ |
| $X_{2,i,2} = 15.475$ | $X_{2,i,6} = 15.900$ |
| $X_{2,i,3} = 15.600$ | $X_{2,i,7} = 16.000$ |
| $X_{2,i,4} = 15.700$ | $X_{2,i,8} = 16.100$ |

*Equation 1(o): Exploratory Oil Wells*

$$Y = 0.059 X_2^{1.043} X_4^{4.691} 0.902^{X_7}$$

where $X_2$ = wellhead price of crude petroleum, adjusted by wholesale prices ($/bbl.)

$X_4$ = production of crude petroleum (billions of bbl.)

$X_7$ = time (year—1939)

$Y$ = AAPG exploratory oil wells (thousands of wells)

Correlation coefficient = 0.9472

Number of observation years = 24

*Equation 2s: South Louisiana Average Wellhead Price Relation*

$$X_8 = a_t X_{1i} + (1 - a_t) X_{2it}$$

where $X_8$ = average wellhead price of natural gas (¢/Mcf) in South Louisiana

$a_1 = 0.060$ ($t$ = year $- 1962 = 1, \ldots, 8$) $\qquad a_5 = 0.305$

$a_2 = 0.120 \qquad\qquad\qquad\qquad\qquad\qquad\quad a_6 = 0.370$

$a_3 = 0.180 \qquad\qquad\qquad\qquad\qquad\qquad\quad a_7 = 0.435$

$a_4 = 0.240 \qquad\qquad\qquad\qquad\qquad\qquad\quad a_8 = 0.500$

$X_{1i}$ = ceiling wellhead price of natural gas (¢/Mcf) in South Louisiana under assumption $i$

$X_{1,1} = 13.7 \qquad\qquad\qquad\qquad\qquad X_{1,4} = 25.7$

$X_{1,2} = 15.7 \qquad\qquad\qquad\qquad\qquad X_{1,5} = 27.7$

$X_{1,3} = 20.7$

$X_{2it}$ = average wellhead price of natural gas (¢/Mcf) in South Louisiana delivered in the year $t$ under contracts effected before 1963, given an effective ceiling $X_{1i}$ on prices under such contracts

$X_{2,1,t} = 13.5$

$X_{2,2,t} = 15.4$

$X_{2,3,t} = 20.2$

For $i = 4, 5$

$X_{2,i,1} = 20.2 \qquad\qquad\qquad\qquad X_{2,i,5} = 20.6$

$X_{2,i,2} = 20.3 \qquad\qquad\qquad\qquad X_{2,i,6} = 20.7$

$X_{2,i,3} = 20.4 \qquad\qquad\qquad\qquad X_{2,i,7} = 20.8$

$X_{2,i,4} = 20.5$

*Equation 3*: *Residential and Commercial*
*Burner-Tip Price Relation*

$$Y_{it} = Y_{1962} + X_{it}$$

where $X_{it}$ = estimated change in national average wellhead price of natural gas (¢/Mcf) under assumption $i$, in year $t$

$Y_{it}$ = residential and commercial burner-tip price (¢/Mcf) under assumption $i$, in year $t$

$Y_{1962}$ = residential and commercial burner-tip price (¢/Mcf) in 1962

(See equation 2 for definition of assumptions)

*Equation 3s*: *Residential and Commercial*
*Burner-Tip Price Relation to South Louisiana*

$$Y_{it} = Y_{1962} + kX_{it}$$

where $X_{it}$ = estimated change in South Louisiana average wellhead price of natural gas (¢/Mcf) under assumption $i$, in year $t$

$k$ = percent of natural gas from South Louisiana consumed in a given state

$Y_{it}$ = residential and commercial burner-tip price (¢/Mcf) under assumption $i$, in year $t$

$Y_{1962}$ = residential and commercial burner-tip price (¢/Mcf) in 1962

(See equation 2s for definition of assumptions)

*Equation 4*: *Industrial Burner-Tip Price Relation*

$$Y_{it} = Y_{1962} + X_{it}$$

where $X_{it}$ = estimated change in national average wellhead price of natural gas (¢/Mcf) under assumption $i$, in year $t$

$Y_{it}$ = industrial burner-tip price (¢/Mcf) under assumption $i$, in year $t$

$Y_{1962}$ = industrial burner-tip price (¢/Mcf) in 1962

(See equation 2 for definition of assumptions)

*Equation 4s*: *Industrial Burner-Tip*
*Price Relation to South Louisiana*

$$Y_{it} = Y_{1962} + kX_{it}$$

where $X_{it}$ = estimated change in South Louisiana average wellhead price of natural gas (¢/Mcf) under assumption $i$, in year $t$

$k$ = percent of natural gas from South Louisiana consumed in a given state

$Y_{it}$ = industrial burner-tip price (¢/Mcf) under assumption $i$, in year $t$

$Y_{1962}$ = industrial burner-tip price (¢/Mcf) in 1962

(See equation 2s for definition of assumptions)

*Equation 5:* *Residential and Commercial Consumption*

$$Y = 0.038 X_9^{-0.894} X_{10}^{0.212} X_{11}^{0.959} X_{12}^{0.608}$$

where  $X_9$ = gas price to residential and commercial consumers (¢/Mcf)
$X_{10}$ = oil price to residential and commercial consumer (¢/gal.)
$X_{11}$ = Number of residential and commercial customers (thousands of customers)
$X_{12}$ = degree days, adjusted
$Y$ = residential and commercial consumption of natural gas (millions of Mcf)

Correlation coefficient = 0.992
Number of states or state combinations = 44

*Equation 6:* *Industrial Gas Consumption*

$$Y = 6355 X_{13}^{2.705} X_{14}^{0.495} X_{15}^{0.812} X_{16}^{0.622} X_{17}^{301} X_{12}^{-0.115}$$

where $X_{13}$ = gas price to industrial consumer (¢/Mcf)
$X_{14}$ = oil price to industrial consumer ($/bbl.)
$X_{15}$ = coal price to industrial consumer ($/ton)
$X_{16}$ = industrial employment (thousands of persons)
$X_{17}$ = steam electric plant power generation (billions of kwh)
$X_{12}$ = degree days
$Y$ = industrial consumption of natural gas (millions of Mcf)

Correlation coefficient = 0.938
Number of States or State Combinations = 42

*Equation 7:* *Production/Consumption Relation*

$$X_3 = 1.601 X_o \left[ \frac{10^{-1}}{7.29 + 87.6(0.7408)^t} \right]$$

where  $X_o$ = consumption of natural gas in year $t$ (billions of Mcf)
$t$ = year—1944
$X_3$ = marketed production of natural gas (billions of Mcf)

*Equation 8:* In the first Southern Louisiana model (which did not assume directionality) this equation was identical to *Permian* Equation 8. In the revised model (taking account of directionality) this equation was omitted.

*Equation 9*:   *Additions to Reserves Relation*

$$Y = 2.595 + 1\,1.472X_{22} + 4.385X_{23}$$

where $X_{22}$ = exploratory gas wells (thousands of wells)
  $X_{23}$ = exploratory oil wells (thousands of wells)
   $Y$ = gross additions to reserves (billions of Mcf)

Correlation coefficient = 0.858
Number of observation years = 16

*Equation 10*:   *Estimated Recoverable Reserves*

$$X_{21,\,t} = X_{21,\,t-1} + X_{20} + X_3$$

where $X_{20}$ = gross additions to reserves of natural gas (billions of Mcf)
  $X_3$ = marketed production of natural gas (billions of Mcf)
 $X_{21,\,t}$ = estimated recoverable reserves of natural gas (billions of Mcf) in year $t$
$X_{21,1963}$ = 277.7 billions of Mcf

*Equation 11*:   *Reserve/Production Ratio*

$$Y = \frac{X_{21,\,t}}{X_3}$$

where

  $X_{21,\,t}$ = estimated recoverable reserves of natural gas (billions of Mcf)
  $X_3$ = marketed production of natural gas (billions of Mcf)
   $Y$ = life index of natural gas (years)

# 6 Model Roles and Presentation Strategy

The purpose of this chapter is to explore the possible roles that an econometric model might assume in the process of regulating natural gas wellhead prices. The answer to this problem, of course, involves the definition of an acceptable operational model. In accomplishing this purpose, consideration is given to the potential degrees of reliance the commission could lawfully place on an econometric model—ranging from a minimum position, in which the model is used only as a justification device, to maximum reliance on such a model, in which estimates made from the model form the guidelines for regulatory decisions.

The preceding chapters have clearly demonstrated that the econometric models developed by the Federal Power Commission's Office of Economics did not directly contribute to the final decisions made in the *Permian* and *Southern Louisiana* cases. In the *Permian* initial decision, the econometric presentation was completely dismissed as being not relevant nor material to the problem. In the *Southern Louisiana* proceeding the econometric models were judged to be not sufficiently reliable to be used as a basis for decision in that hearing; but in his written opinion the presiding examiner stated that the econometric approach was worthy of continued pursuit. This limited recognition given to the econometric analysis in the *Southern Louisiana* proceeding has been a major stimulus to the Office of Economics to increase its efforts toward the development of a more suitable and reliable econometric approach.

The failure of the econometric presentation in the *Permian* proceeding, under slightly different circumstances could have brought about the complete rejection of further econometric experimentation as applied to Federal Power Commission regulatory matters. This assertion is supported by examining the problem dealt with in the *Permian* case; the *Permian* hearing was the first proceeding whose purpose was to determine just and reasonable area rates and the proceeding was to serve as a model for future area rate procedures.[1] The most qualified staff personnel were assigned to the *Permian* case and the hearing examiner permitted almost unlimited time periods to be taken in the introduction of testimony and for cross-examination. The large investment of time and talent devoted to the *Permian* proceeding was made to develop a precedent for future area rate hearings. If the second area rate hearing had not started while the first area rate proceeding was still in progress, the desirability of presenting essentially the same econometric study for the second proceeding might have been seriously questioned. The prepared econometric testimony for the *Southern Louisiana* models was virtually identical to that used in the *Permian* case; it is doubtful that this analysis would have been presented by the staff if the examiner's decision had been given just prior to the scheduled econometric testimony offered in the *Southern Louisiana* case. On the other hand, if these rate hearings had been separated by a period of several years, it might have been possible to develop a much improved and more successful model for the second area rate proceeding, but this view assumes that there would have been a sufficiently strong interest within the commission to support more extensive econometric research. The published documents pertaining to the *Permian* hearings and the Federal Power Commission publications in general do not imply that the *Permian* experiment in econometrics displayed sufficient merit to warrant the continuation of this type of analysis.

The second econometric exercise in natural gas wellhead price regulation was in the record slightly before the decision was final in that of the first; but, even though the econometric studies were nearly identical, the *Southern Louisiana* presentation was received more sympathetically than was the *Permian*. Neither presentation significantly influenced the examiner's decision; but the *Southern Louisiana* econometric analysis suggested a worthwhile potential, which had not been officially recognized earlier. No specific reasons were given as to why the second econometric presentation elicited more favorable comment and encouragement than the first, but careful reading of the examiner's decisions suggests that the econometric presentation and defense in the *Southern Louisiana* might have won its limited praise because it was less pretentious and more candid in recognizing the limitations of the analysis presented. In the *Southern Louisiana* proceeding, the econometric approach defended by the staff was somewhat more flexible than it had been before, and perhaps the introduction of this pliability suggested to the examiner a more promising future role for such models in the regulatory process. It can be asserted that the staff's econometric approach gained its first serious recognition in the *Southern Louisiana* proceeding. Since the commission appears to be willing to encourage the staff to develop further econometric models to assist in future area rate proceedings, it is important to evaluate the strategies of model presentation and to comment on the variety of functions that such models might perform.

For the policy maker, econometric models are devices of prediction. Those who develop models and submit them as decision-influencing tools in regulatory proceedings must avoid the temptation of enveloping their work in an aura of verity. It must be very disturbing to a hearing examiner to learn that an econometric model—which was first presented as a scientific predictive technique that would remove much of the stigma of a priori and ad hoc reasoning from regulatory proceedings—may in essence be more vulnerable to attack than the regulatory analytical procedures used in the past. Those analysts who develop econometric models to be used in determining policy must resist the temptation to oversell the objective capabilities of their creations and, instead, stress the rationality of their econometric approach as being no more than a more precise, logical extension of the traditional approaches. The econometrician must demonstrate that in developing his model he has attempted to take into account the established goals of policy, the traditional variables of regulation, the legal constraints that protect and inhibit the interested parties, and all of the available data the model builder has selected to describe the area subject to regulation. The model builder should explain that he has attempted to simplify the environment he is describing by removing from his model all of those elements present in the real environment that do not significantly influence the primary factors under consideration. In short, he must stress that he is using strategic variables. The elements that remain after the elimination of the less relevant variables are those that constitute the model.

The econometrician begins his work, therefore, just as any other analyst; but, as he proceeds, he stresses statistical relationships where possible and employs "quantified expertise" when the tools of statistical analysis or the availability of data are not sufficient. In one sense, the econometric model is more objective than traditional analysis; in another sense, it is more subjective. The model is more objective in that it is a complete, mathematically precise statement of the problem solution; it is more subjective in that its a priori assumptions may be more obscured, and the statistical and mathematical processes employed to derive its relationships may not be readily discernible. That is, from a superficial point of view, an econometric model may exhibit all of the characteristics of scientific objectivity even though many of its essential components were developed subjectively.

The econometric models presented in future regulatory hearings will have to be developed in part through highly subjective processes. There are many points reached in the course of the analytical process of developing an econometric model at which the analyst must choose from among competing alternatives with respect to basic structural assumptions, data categories and sources, statistical techniques, and other similar items. The analyst must make such decisions by relying on his best judgment or available advice as his model progresses from its initial conceptual stages toward its final operational applications. Econometricians who seek to develop models as policy tools should be especially aware of the consequences of choosing one approach over a second; and, during the model development process, a formal justification should be recorded for each major choice. An analyst who is attempting to develop a model worthy of policy application should go to great lengths to evaluate the end effects of incorporating selected alternative choices in a given model; this point is especially important when it is known that the model will be challenged in adversary proceedings. Given sufficient time and qualified personnel, the Office of Economics should undertake a methodological study that will focus on the specific points or places in the construction of the model where significant decisions have to be made. The model builders should know in advance what options or alternatives they must select from, and at what stage or time in model development the choices will have to be made. These choices will be made from options concerning statistical techniques, economic theories, model structures, industrial factors, data sources, and other similar factors. If the analyst is consciously aware of the many alternatives from which he must choose and if the scheduling or timing of these decisions is anticipated, the model can be developed in a more consistent and honest fashion. Moreover, attacks in adversary proceedings can then be anticipated.

A very desirable approach to model development is one involving the complete evaluation of the feasible alternatives so that the ultimate effects of specific decisions can be better evaluated. For example, if at a given point in model development, there are two statistical techniques for developing an estimating relationship, both methods should be used, and the resulting differences in estimates resulting from the two different methods should be evaluated. If the relationships and the estimates are substantially the same, the analysts can have a higher degree of confidence in their work than if the relationships have produced widely conflicting estimates. It is quite obvious that when a choice is made between such alternatives and some portion of the validity of the model depends on that choice, the model builder must evaluate both alternatives if he is to have confidence in his model. The Office of Economics should attempt to develop future econometric models by means of the evaluation of alternatives so that the staff would be in a better position to present and defend its analytical procedures. It is not possible to require that all alternatives be tested, but it is necessary to evaluate the effects of the more important choices.

The strategy of presentation of staff econometric testimony in regulatory procedures should be carefully developed to insure effective model utilization. Staff witnesses might begin their testimony by raising an issue not found in the earlier area rate hearings; this issue is whether the natural gas industry can be described through the vehicle of an econometric model. Further, if the industry can be described, can the descriptive mechanism also be useful for prediction?

This fundamental question should be raised prior to the assertion that the staff has a valid model, which describes and predicts. The experts in econometrics who testified for the natural gas industry in the previous area rate proceeding did not comment on the suitability of econometric analysis for cases of this type, but one would infer from their testimony that they considered the econometric approach applicable. Therefore, if the staff begins its econometric argument with the contention that the econometric methodology is applicable in such cases, it will focus its initial argument on a fundamental and vital point. It would be difficult for an established expert in econometrics to testify for the gas interests that his specialty does not apply to the economics of regulation; it should be recalled that experts in the previous hearings (Drs. Barton and Cootner, *et al.*) discredited staff's models primarily by asserting their personal technical superiority in econometrics. In future hearings, if it is established that an econometric model can make a contribution to the regulatory procedure, the staff will be able to introduce its model with more authority.

In future hearings, the staff's econometric model should be presented in a very general sense when it is first introduced. The precision and complexity of the model should not be featured in its first exposure. Instead, the staff should explain the concept and structure of its model and comment on its best general estimates of coefficients for the variables in the model. This early testimony should point out whether the coefficients were positive or negative and whether their values were relatively large or small rather than introducing the complete, complex model with its coefficients exactly carried out to many decimal places. The staff econometric presentation should follow a sequence that begins with the discussion of the basic validity of the econometric technique, progresses to a general description of the staff model, and then concludes with the broad findings and recommendation drawn from the model. This format for initial presentation would not appear to be as vulnerable to the type of attack aimed at previous econometric models. If the staff has argued that econometrics can be usefully employed in the regulatory process, that its experts have used this approach, and that some general findings have been made, this sequence of testimony may require a more positive type of rebuttal by industry experts in which alternative estimating relationships, submodels, or models are offered in opposition to the staff's presentation. If the Office of Economics had adequately prepared its brief describing the development of its model, the staff should be in good position to evaluate the econometric alternatives proposed by industry experts. These alternatives should not greatly surprise staff analysts if they have built their own model through a sequential process of evaluating alternatives. The staff strategy of econometric presentation should be one of maintaining an initial focus on primary issues so that some fundamental agreements are reached prior to the inevitable arguments that concern more peripheral points.

There is a wide range in the degrees of reliance the examiner and ultimately the commission might place on an econometric model. Examples of the case of no reliance can be found in the *Permian* and *Southern Louisiana* hearings. In each of these hearings, the examiner permitted long periods to be devoted to the development of the staff's econometric presentation and to industry's cross-examination; but in each case, the examiner's decision was not directly influenced by this testimony. If many more hearings are held in which staff models are judged to be not applicable, econometric presentations will undoubtedly be discontinued.

There are several possible contributory roles that models might play in future proceedings. One possibility might be the use of a model to describe and help justify a decision after it has been substantially determined. In such a case the econometric model would be developed to describe the given position or conclusion reached by the hearing examiner. This type of *ex post* model building would not be undertaken to provide proof or support for the examiner's decision; it would be constructed to illustrate the examiner's decision-making process thereby making the process of the decision more explicit. This type of application of an econometric model could be very useful in the formulation of a given regulatory decision because it would help the hearing examiner by objectively reviewing his logical sequences and by illustrating how a model developed in this *ex post* fashion would function with slightly altered assumptions, different parametric inputs, or in future cases. It is quite possible that hearing examiners who helped formulate econometric models of this type might specify that such models were to be considered unique to a given case and not applicable to future proceedings. This qualification would recognize the *ex post* nature of model development in its admission that a rather " nonscientific approach " had been taken to model building. This type of development of an econometric analysis of natural gas regulation would provide a minimum use of a model, but from this limited, *ex post* approach a more significant contribution might develop.

Greater reliance on an econometric approach might follow from this use of *ex post* models in several cases. This result could occur as hearing examiners become more familiar with the econometric approach and as econometric models from previous hearings were employed as current references. Through such a process, the econometric approach could slowly mature and become a more determinate tool of regulation. As the staff and the advocates of the gas interests become more familiar with the use of econometrics in regulatory proceedings, models will assist in providing a more reasonable approach for regulatory analyses. Reasonableness, as used in this sense, is that condition which produces a result sufficiently reliable in the eyes of the competing interests involved so as to provide a basis for further action. This approach involves acceptance of an instrumentalist approach or adoption of a procedure of self-correcting value judgments. This role is the proper one for econometric analysis as applied to policy problems in an adversary proceeding.

It is conceivable that the hearing examiner could place primary reliance on an econometric model developed by the staff. In such a case the model would assume a role of maximum importance and the examiner's decision would be based upon model estimates. This was essentially the role sought by the staff for the econometric model of the *Permian* hearing. One of the major reasons why the *Permian* model was not successful in helping to shape the examiner's decision was that the staff broadly overstated the role and application of its model early in the proceedings. Even if an examiner were convinced by staff econometric arguments and his decision were based on testimony derived from model predictions, it is questionable that such decisions based on model evidence would be upheld in courts of appeal. It can be concluded that, at this time, there is little chance for success of an econometric presentation by staff, which is designed to introduce a model that will assume a determinate role in regulatory proceedings and supplant the necessity for judgment by the hearing examiner.

This chapter has considered the various roles that an acceptable econometric model might play in the regulation of natural gas wellhead prices. Concurrent with the evaluation of potential roles, a staff strategy was suggested for presentation and development of model testimony. Following this discussion of presentation format an analysis was developed which led to the assertion that models used in regulatory proceedings should be developed through a reasonable and instrumentalist approach. The following chapter examines the relationship between due process of law and the potential uses of econometric models by regulatory commissions such as the Federal Power Commission. Consideration is also given, in the next chapter, to the possibilities that the econometric technique as a tool for regulation may spill over into the kit of tools of other regulatory commissions.

# 7

## Use of An Econometric Model and the Problem of Procedural Due Process of Law

The purpose of this chapter is to evaluate the problems that might arise from the use in a regulatory proceeding of an econometric model with respect to the constitutional guarantee of due process of law. This vital constitutional issue may arise if econometric models are substantively used in future rate proceedings to supplant commission judgment. It is therefore important to examine carefully the relationship between due process of law and the development and application of econometric models by regulatory commissions.

It is not easy to give a short and simple definition of the phrase, "due process of law." Applicability of this concept to regulatory agencies has, however, been asserted in general terms by John R. Commons as follows:

... the history of Anglo-American jurisprudence is a history of efforts to work out fundamental principles of classification which shall permit new proportioning of the national economy without unduly disturbing the old. This history is epitomized in the largest term known to jurisprudence, "due process of law" ... the officers of government are limited by due process and equal protection, but within those limits they may reproportion the national economy by a reclassification of persons for the purpose of assigning to them what is deemed a proper share in the expected burdens and benefits of the commonwealth.[1]

Justice Roberts in *Nebbia* v. *New York* noted that the courts must be concerned with the procedure of due process and not with "substance" when he asserted:

The Fifth Amendment, in the field of federal activity, and the Fourteenth, as respects state action, do not prohibit governmental regulation for the public welfare. They merely condition the exertion of the admitted power, by securing that the end shall be accomplished by means consistent with due process. And the guarantee of due process ... demands only that the law shall not be unreasonable, arbitrary or capricious, and that the means selected shall have a real and substantial relation to the object sought to be obtained.[2]

The quotation from Commons given above refers to his recognition of the importance of the concept of procedural due process with respect to questions involving private rights versus public needs, and Justice Roberts' citation spells out one current procedural requirement of due process. Commons explained his position in greater detail by saying also:

... the process of classification and reclassification according to the purposes of the ruling authorities [is] a process which has advanced with every change in economic evolution and every change in feelings and habits toward human beings, and which is but the proportioning and reproportioning of inducements to willing and unwilling persons, according to what is believed to be the degree of desired reciprocity between them. For classification is the selection of a certain factor deemed to be a limiting factor, and enlarging the field of that factor by restraining the field of other limiting factors, in order to accomplish what is deemed to be the largest total result from all. . . . In so far as procedure is deemed necessary for these general purposes it resolves itself into that minimum of procedure by which all of the facts are brought before the court, including opportunity of the defendant to be heard through counsel, with all that implies of notice and approved judicial methods of investigation.[3]

Due process of law with respect to regulatory commissions primarily involves observing a procedure for following a specific sequence of steps.

The first step in the sequence required for due process is that adequate notice must be given to the interested parties that a public hearing is to be held. The Federal Administrative Procedure Act defines this requirement of notice as follows:

Persons entitled to notice of an agency hearing shall be timely informed of (1) the time, place, and nature thereof; (2) the legal authority and jurisdiction under which the hearing is to be held; and (3) the matters of fact and law asserted.[4]

This notice must include a statement informing the interested parties of the specific issues that will be involved in the hearing. The second step in this sequence is a *public hearing*, which gives the parties involved an opportunity to introduce evidence and testimony in support of their interests. The Federal Administrative Procedure Act defines this procedure as:

The agency shall afford all interested parties opportunity for the submission and consideration of facts, arguments, offers of settlement, or proposals of adjustment where time, the nature of the proceeding, and the public interest permit.[5]

The hearing permits the presentation of evidence, cross-examination of opposition witnesses, and rebuttal testimony. The third step is the final commission decision which is made following the conclusion of the public hearing. In the case of the Federal Power Commission, an initial decision is first made by a hearing examiner and copies of this decision are sent to all interested parties. If no objections are filed to this initial decision within thirty days, this decision then becomes the final decision of the commission. Decisions issued by the hearing examiner and adopted by the commission must set forth *findings of facts*. These findings are the conclusions reached on disputed factual issues; the Federal Administrative Procedure Act requires that such commission findings be supported by substantial evidence upon consideration of the whole record. The final step in this sequence is that of availability of judicial review. Judicial review is undertaken only after all commission remedies have been exhausted. The most common grounds for such review have included the following:

1. The statute relied upon by the commission was unconstitutional,
2. The commission exceeded its statutory authority or jurisdiction,
3. The commission's order was not supported by findings,
4. The findings were not supported by substantial evidence,
5. The order violates specific constitutional guarantees,
6. The order resulted from unlawful procedure or some other error of law,
7. The order was based on a misinterpretation of the law administered by the commission.[6]

The Supreme Court has held that judicial review of commission decisions should focus on questions of power and right and not on issues that can best be resolved through the administrative discretion of a commission. The Supreme Court asserted this point in a 1910 Interstate Commerce Commission case by stating:

Beyond controversy, in determining whether an order of the Commission shall be suspended or set aside, we must consider: a, all relevant questions of constitutional power or right; b, all pertinent questions as to whether the administrative order is within the scope of the delegated authority . . .; c, whether, even although the order be in form within the delegated power, nevertheless, it must be treated as not embraced therein, because the exertion of authority which is questioned has been manifested in an unreasonable manner.[7]

In a later case the Supreme Court summarized its position on its role in the review of commission orders by asserting:

Once a fair hearing has been given, proper findings made and other statutory requirements satisfied, the courts cannot intervene in the absence of a clear showing that the limits of due process have been overstepped. If the commission's order as applied to the facts before it and viewed in its entirety, produces no arbitrary result, our inquiry is at an end.[8]

From decisions such as these, it is clear that the courts have usually held that a finding of fact by a regulatory commission is conclusive if it is based on substantial evidence.

A typical example of the issue of due process of law as raised in a case involving a federal regulatory decision is found in the recent Supreme Court opinion delivered by Justice Fortas in the *Penn-Central Merger and N and W Inclusion Cases.*[9] In the opinion—which reviewed a decision and order made by the Interstate Commerce Commission—the Court denied an appeal from the commission order after finding that the commission had followed the procedure required by due process of law. This opinion, in tracing through the required sequence of steps, first established that the merger and inclusion orders were within the established congressional policy of encouraging railroad mergers as formulated in the Transportation Act of 1920 and the later Transportation Act of 1940. The opinion next indicated that exhaustive hearings had been held in which approximately 200 parties had participated—including the commission's own staff and representatives from the Department of Justice. The Court next found that the commission's decision had reasonably balanced public and private interests within the framework of the governing statute; Justice Fortas concluded by stating:

Considering the record, and the findings and the analysis of the Commission, we see no basis for reversal of the District Court's decision that the Commission's "public interest" conclusions are adequately supported and are in accordance with law. We find no basis, consonant with the principles governing judicial review, for setting aside the Commission's determination . . . that the directives of the governing statute have been reasonably satisfied; that the transaction is likely to have beneficial and not an adverse effect upon transportation service to the public; and that appropriate provisions have been made with respect to other railroads that are directly affected by the merger.[10]

The expertise of the Interstate Commerce Commission which was employed to determine the value and exchange ratio for inclusion of certain railroads in the merger was attacked by individual parties to the merger. Some of the interested parties asserted that inclusion prices were too high; other parties charged that the inclusion prices were too low; and it was argued that some aspects of the inclusion formula were arbitrary. Concerning these objections Justice Fortas noted:

The method for determining the value and exchange ratio which the Commission adopted . . . is a method that is reasonably conventional and generally accepted, always subject to the modifications and adaptations required by individual cases, and we see no basis for holding it erroneous as a matter of law. The attack that is launched is upon factors of particularized judgment and the weight to be ascribed to various values. These are matters as to which reasonable men may differ in detail, and we see no basis for setting aside the Commission's conclusions as sustained by the District Court. In setting inclusion terms, the Commission was dealing with complicated and elusive predictions . . . we are no more competent than the Commission and the District Court to ascertain the accuracy of these predictions. We deem it our function, in the complexities of a case such as this one, to review . . . with respect to agency actions to make certain that those actions are based upon substantial evidence and to guard against the possibility of gross error or unfairness. If we find those conclusions to be equitable and rational, it is not for us to second-guess each step in the Commission's process of deliberation.[11]

The above excerpt from Justice Fortas' opinion indicates the traditional reluctance of the Supreme Court to attempt to evaluate a commission's weighting techniques and analytical methods that shape individual points in a given commission decision and subsequent order.

How does an econometric model which is used in rate regulation fit into the issue of due process of law? If the model is used by a commission to evaluate alternative rate policies, the model is then an instrument for pro-portioning private rights and public interests. Due process of law demands that the regulatory process shall not be unreasonable, arbitrary, or capricious, and the question of the use of an operational econometric model must be carefully investigated to insure that that use is not unreasonable, arbitrary, or capricious. If models are developed for specific rate hearings and if those models are instrumental to the decision-making processes of regulation, an uncritical acceptance of such a model by the commission may be interpreted to constitute a capricious finding of fact. If the hearing examiner states that he feels that an econometric model is sufficiently reliable to be used in evaluating the consequences of a set of proposed area prices, this finding constitutes a finding of fact. If a commission decision in which a model had been employed is reviewed by a court, an interesting question may develop as to just how much of the given econometric model is to be considered a finding of fact. Is the entire model, as used, to be considered a finding of fact? If the complete model is not found as a fact, what portions of the model are to be so found? The model's structure, variables, and coefficients might all or in part be considered to be findings of fact by a commission and outside the normal range of court review upon appeal. If a model and its principal components are treated as findings of fact, this action will tend to reduce the role of the hearing examiner in regulatory matters and strengthen that of the economic staffs of commissions. Regulation by commissions which rely on econometric models may bring fundamental changes that were not antici-pated by the first model builders. Heavy reliance on econometric models would constitute, in effect, a situation in which commissioners would abandon major portions of their decision-making power to the model builders.

The relationship between due process and regulatory econometric models will be determined by the attitude of the courts toward a commission's use of fact and associated expertise as formulated in a given model. In its review of Federal Power Commission decisions in the past, the Supreme Court has followed the established pattern of accepting commission rulings and findings of fact which were supported by substantial evidence in the hearing record. In the *Hope Natural Gas Case* the majority decision asserted:

We are not obliged to examine each detail of the Commission's decision; if the total effect of the rate order cannot be said to be unjust and unreasonable, judicial inquiry under the (National Gas) Act is at an end.[12]

Justice Harlan amplified the point in his delivery of the *Permian Basin Area Rate Cases* opinion by stating:

A presumption of validity . . . is attached to each exercise of the Commission's expertise. . . . Moreover, the court has often acknowledged that the Commission is not required by the Constitution or the Natural Gas Act to adopt as just and reasonable any particular rate level; rather, courts are without authority to set aside any rate selected by the Commission which is within a "zone of reasonableness." No other rule would be consonant with the broad responsibilities given to the Commission by Congress; it must be free, within the limitations imposed by pertinent constitutional and statutory commands, and to devise methods of balancing conflicting interests. It is on these premises that we proceed to assess the Commission's orders.[13]

It should be safe to assert, with respect to the above quotations, that the Supreme Court would not rule against the use of an econometric model as a commission regulatory device. As a method or a tool of analysis the basic econometric approach would undoubtedly be an acceptable commission technique as long as the model had been developed in a manner that did not violate procedural due process of law. It would then be difficult, however, to argue that a regulatory commission had observed all of the requirements of due process of law in reaching a given decision, especially if the examiner had relied on an econometric model that had been developed by the commission's economics staff prior to the specific hearing. The purpose of the hearing is to give all interested parties an opportunity to introduce evidence, cross-examine, and relate opposing evidence. If a commission staff introduces an econometric model into a hearing as a *fait accompli*, the essential purpose of such a hearing cannot be realized. An econometric model that has been developed by a single group to describe and predict the most significant parameters that influence the market activities of an industry subject to regulation always includes some subjective characteristics. The procedure of a fair hearing must permit the identification and reevaluation of those portions of an econometric model which are challenged by interested parties. When a complete econometric model is introduced in a hearing, it should be treated as though it is a hostile witness by those who have not had a hand in its development. Interested parties must have the opportunity and sufficient time to evaluate the original development and ultimate impact of each component part of a given econometric model. During the process of the hearing, certain portions of the original model will be validated while others will be successfully challenged. The development of a viable econometric model which is to be used as a regulatory device must be accomplished, in part, during the very processes of the scheduled hearing itself. The model as a finding of fact must be based on substantial evidence as presented by parties participating in the hearing if its use is to conform to the constititutional guarantee of procedural due process.

The use of econometric models in regulatory proceedings need not be restricted to hearings conducted by the Federal Power Commission. This technique can be adapted to help other commissions perform their regulatory functions. This tool of econometrics would be especially helpful when applied to a regulatory environment somewhat analogous to that typical of natural gas production. Regulation of price within an industry composed of firms whose costs may vary widely may be somewhat simplified by judicious use of the econometric methodology.

There are five federal commissions with jurisdiction over interstate transportation and utility services. The Federal Power Commission is one of these agencies, and its attempts at employing the econometric methodology for rate analysis have been the focus of this study. How suitable would the econometric approach be as a major tool of rate analysis for the other four commissions? The Interstate Commerce Commission, which has jurisdiction over railroads, oil pipelines, interstate motor and water carriers, freight forwarders, and express companies, should seriously investigate the possible applications of econometric models to aid in the processes of establishing just and reasonable rates and associated commission orders, particularly in cases involving group rates. The Federal Communications Commission, which took over the regulation of interstate telephone and telegraph services from the Interstate Commerce Commission and assumed control over radio and television from its predecessor, the Federal Radio Commission, could also find areas in which econometrics might profitably be applied. The Civil Aeronautics Board, which has economic control over commercial air transportation, should also find some potential uses for econometrics in its rate procedures. The remaining federal commission, the Securities and Exchange Commission, does not now have a primary rate-making responsibility, and, therefore, this commission would have the least current interest in the econometric techniques of rate making.

The pioneering efforts of the Federal Power Commission in the investigation and attempted application of econometric models may provide a stimulus for other commissions to undertake similar analyses. The strength of this stimulus will, in part, depend on the degree of success the Federal Power Commission's economics staff achieves in future attempts to employ econometric models. Spillover from this general economic technique into the kit of tools of other regulatory commissions will be much more likely if the Federal Power Commissioners in the future place more reliance on the econometric approach than they have in the past.

This chapter has examined the relationship between the uses of an econometric model by a regulatory commission and the constitutional guarantee of due process of law. In achieving this end the meaning of procedural due process was first developed. After defining this concept, certain issues relating to the question of whether or not an econometric model is a finding of fact were examined with respect to the guarantees of due process. It has been concluded in this chapter that an econometric model can be used in rate hearings so long as the model's most sensitive components are essentially developed during the hearing from substantial evidence introduced into the hearing record. This chapter has indicated a doubt that an econometric model developed by a staff prior to a hearing can be used as a regulatory device without potentially violating the protective guarantees of due process of law. This chapter has considered the possible use of this type of econometric methodology by other federal regulatory agencies, and it has been suggested that other agencies concerned with rate making may find this technique useful. The following chapter concludes this study. It will review the study's purposes, its method, and the conclusions and recommendations which have been developed in the process of completion of this analytical inquiry.

# 8 Conclusions and Recommendations

The purpose of this study, as identified in Chapter 1 has been to develop a critical evaluation of the uses of an econometric model in cases involving natural gas rates. This evaluation has focused on the specific econometric approach to price regulation developed by the Federal Power Commission's Office of Economics for the *Permian Area Rate Hearings*; this model, in slightly revised form, was also introduced in the *Southern Louisiana Area Rate Hearing*. A secondary purpose of this study has been to analyze the question of the potential usefulness of econometric models as tools of analysis in rate proceedings before other regulatory commissions. These purposes have been accomplished within the preceding chapters by: (1) an examination of the circumstances that led to the development of the basic model used in the *Permian Area Rate* case, (2) a comprehensive review and analysis of this model's first actual application in the *Permian* proceeding and its second application in the *Southern Louisiana Area Rate Hearing*, (3) a critical examination of the basic assumptions and techniques employed in structuring the components of these two econometric models, (4) the suggestion of a set of revisions of these models that might make them more applicable to regulatory hearings, and (5) the recommendation of a radically different strategy from that employed in these cases for the development and employment of econometric models in future rate hearings.

This final chapter will provide a concise summary of the material developed in the preceding pages and conclusions based on the analysis in preceding chapters. Chapter 1's introduction to the investigation at hand included a detailed account of the purpose, nature, scope, and organization of this study. These concluding pages are, therefore, devoted first to a review of the most important findings made within the chapters and, next, this writer's recommendations as to what might improve the effectiveness of the use of econometric methodology as a regulatory tool.

## Conclusions

The Federal Power Commission has been placed in the position of acting as the regulator of wellhead prices in the natural gas industry. This regulatory responsibility was not sought by the Federal Power Commission, but was thrust on the commission by the Supreme Court in the decision reached in the *Phillips Case*.[1] The requirement that the Federal Power Commission must establish wellhead natural gas rates created many difficulties which, the commission judged, could not be satisfactorily handled by traditional regulatory means. The usual cost-of-service approach to regulation was not applicable to natural gas wellhead rates because of (1) the problems associated with the allocation of joint costs for fields and wells that produce both natural gas and crude petroleum, (2) the lack of a consistent relationship in the natural gas exploration industry between cost of discovery and the value of the product discovered, and (3) the administrative complexity of attempting to deal with a very large number of producers on an individual basis. The commission's temporary solution to these problems was to issue an interim set of wellhead price guidelines to apply to twenty-three pricing areas. Through this area pricing approach—which combined all producers in a given field into a common rate grouping—the commission met its responsibility of setting wellhead rates, but the first set of rates (which were issued in 1960) were merely interim prices, which it was necessary later to replace by a system of area prices that could be defended as being "just and reasonable" under the Federal Power Act. The *Permian Basin Area Rate Proceeding* was the first proceeding having stated a purpose of determining "just and reasonable" rates for natural gas for a specific area. The Office of Economics within the Federal Power Commission developed an econometric model to be used for this purpose, and this model was introduced in the *Permian Area Rate* case for the purpose of providing the commission with an efficient and flexible tool to deal with some of the most difficult problems of gas rate regulation.

The econometric model developed for the *Permian* hearing was designed to project the effects that a selected range of wellhead gas prices would have on the supply of and demand for natural gas. As conceived, this model used as its inputs an assortment of hypothetical gas prices and, in turn, projected estimates of future exploratory effort, of additions to gas reserves, of gas production, of gas consumption, and of other factors that might affect the natural gas industry and the related public interest. The *Permian Basin* model consisted of three separate parts; these three components separately treated exploration activity, residential–commercial consumption, and industrial consumption. The basic analytical approach used in developing the relationships selected as the most representative of each of these three areas was that of regression analysis. The model was composed of three submodels—one submodel for supply (exploratory activities), one for demand, and a "linkage" submodel which united the first two. The *Permian* econometric model was a "closed model" because it contained provision for a feedback relationship that made the final output in the demand submodel one of the influencing factors in the first equation of the supply submodel.

The most important general findings derived from this method by its developer were (1) that past increases in wellhead prices had not resulted in additional new gas reserves and (2) that any future price increases would only lessen exploration and thereby reduce new additions to reserves. The model builders of the Office of Economics explained the known increase in gas sales and reserves over the past decades by pointing to the rapid growth of the economy and the displacement of other fuels. The model builders argued that these growth factors had largely outweighed the depressing effects of price increases up to the present, but now, the model builders asserted, these growth forces are less vigorous than before and also gas prices are at a level where competition from other fuels is being felt. Those who had developed the model argued that the only way in which the gas industry could achieve a favorable growth rate in future years would be to reduce gas prices from their 1961 levels.

Expert witnesses, representing the natural gas interests, challenged virtually every component of the commission staff's econometric model. It was established that these witnesses were individually technically better qualified to discuss the natural gas industry than were the commission's model builders in every academic and industrial aspect relevant to the model and its related hearing proceedings. The natural gas industry's experts charged that the model's authors were guilty of the following: false sophistication, non-professional performance, faulty use of data, incorrect identification of variables, statistical ineptitude, and conceptual inconsistency. The expert witnesses for the industry were able to support their criticisms to the satisfaction of the hearing examiner. Two specific points, which developed from the above list of broad charges, were found to be most damaging to complete reliance on use of the model. These points were, first, that those who had built the model had not made any positive provision in the model for the directional capability of exploratory activities and, second, that the model presented in the *Permian* case was a national model, which could not be uniquely related to the demand for and supply of gas relative to *Permian* area producers.

A second area rate proceeding was initiated prior to the completion of the first. The *Southern Louisiana Area Rate Hearing* was nearing its closing stages when the Presiding Examiner's Initial Decision on the *Permian* hearing was published. The Federal Power Commission's Office of Economics placed essentially the same econometric model that had been used in the *Permian* hearing in evidence in the *Southern Louisiana* case. This econometric model as first introduced into the *Southern Louisiana* case differed only slightly from the *Permian* model in that the *Southern Louisiana* model contained an alternate submodel for making demand projections for the specific area involved in the hearing. The model builders did not develop a separate submodel for estimating areas gas supply or exploratory activities because they assumed that the intensity of exploratory activities could best be explained on the national level. The *Southern Louisiana* demand submodel employed the basic national consumption equation, as used in the *Permian* model, and output values from this equation were adjusted to provide an area estimate by multiplying each state's projected consumption by the fractional amount that the *Southern Louisiana* area had supplied to the state in 1961. This method of relating the national demand submodel to the specific area was developed by use of a very rudimentary technique.

The econometric model introduced in the *Southern Louisiana* hearing had included a regional submodel because, during the cross-examination of the commission's witnesses in the *Permian* hearing, it was obvious that the inability of the model builders to relate their national projections to the area subject to regulation was a serious shortcoming. Shortly after the *Southern Louisiana* model had been introduced, Federal Power Commission Opinion 468 was issued. This opinion was the commission's decision respecting the *Permian Basin* proceeding. In this opinion, the commission specifically stated that the staff's *Permian* econometric model was not applicable to the rate proceeding because it had not included a provision for directionality of exploration. Immediately following the issuance of this opinion, the *Southern Louisiana* hearing examiner asked that the econometric model offered in evidence on the latter case be revised to include a directional capability. Following his request, the model builders hastily revised their work and re-introduced a model that assumed perfect directionality over the entire twenty-four year period from which the model's equations were derived. This assumption of perfect directionality, however, was found to be grossly inaccurate because directional capability was clearly a recent phenomenon. It can be concluded that the most serious faults found in the *Southern Louisiana* model were the same as those of the *Permian*; the models did not realistically relate to the *specific* areas subject to regulation, and the models did not have a workable mechanism to deal with the recently successful directional capabilities of the oil and gas production industry.

It would not be correct to asume that the econometric presentation given in the *Southern Louisiana* hearing was as spectacular in its failure as it had been in the *Permian* proceeding. The performance of the individual who introduced the *Permian* econometric model was, in large part, responsible for the summary dismissal of the model by the examiner. The *Permian* model was confidently introduced with assertions that it would be a powerful new tool for analyzing the most difficult aspects of the complexities of natural gas rate regulation. When the positive claims for this model were challenged by expert witnesses, who were employed by the gas interests, the individual primarily responsible for the model was not able or willing to reconcile his initial claims for the use of the model with its subsequently demonstrated inadequacies. In contrast, the individual who introduced the econometric model into the *Southern Louisiana* hearing employed a rather low-key, more flexible approach —which was highly praised by the hearing examiner. Therefore, even though the econometric model did not influence the decision in the *Southern Louisiana* hearing, the econometric presentation did succeed in suggesting to the examiner that, with more refinement, econometric models could prove to be valuable tools in future regulatory proceedings. The second econometric presentation elicited more favorable comment and encouragement than the first because it was less pretentiously introduced and its supporters more candidly recognized the limitations of their analysis.

### Recommendations

For the policymaker, econometric models are devices of prediction. Those who develop models and submit them as decision-influencing tools in regulatory proceedings must avoid the temptation of enveloping their work in an aura of verity. Analysts must also avoid the temptation to oversell the objective capabilities of models, and should, instead, stress the fundamental rationality of the econometric approach. The econometrician must demonstrate that in developing his regulatory model he has taken into account the established goals of policy, the traditional variables of regulation, the legal restraints that protect and inhibit parties, and all of the available data pertinent to the issues that constitute the regulatory hearing.

One of the most important recommendations this study can make concerns the method of introduction of an econometric model into a regulatory proceeding. The presentation of a model is important, first, as a point of strategy and, second, as a point of law. The strategy of presentation of econometric testimony should be carefully developed to insure effective model utilization. Commission staff witnesses should begin by raising the issue of the applicability of the econometric technique as a vehicle for describing the gas industry. If it can be established that this industry can be described by an econometric model, the question of the predictive capability of such a descriptive mechanism should be raised. These issues are fundamental and should, therefore, be examined prior to the assertion that the staff has a valid model that can be used to describe and predict.

The staff's econometric models should, in the future, be presented in very general terms during initial testimony. Precision and complexity should not be a major feature of first exposure. Emphasis should first be placed on model concept and structure, and this should be followed by a discussion of best initial estimates of the coefficients of the variables in the model. This early testimony should point out whether coefficients, as estimated, were positive or negative and whether their values were relatively large or small rather than attempt to introduce complete, complex models with coefficients precisely stated to many decimal places. The staff econometric presentation should follow a sequence that begins with a discussion of the basic validity of the econometric technique, progresses to a general description of the proposed model, and then concludes with the broad findings and recommendations drawn from the model. This format for initial presentation would not appear to be as vulnerable to the type of attack that was aimed at previous econometric models. If a commission's staff argues that econometrics can be usefully employed in the regulatory process, that its experts have used this approach, and that some general findings have been made, this sequence of testimony may require industry experts to offer a more positive type of rebuttal in which alternative estimating relationships, submodels, or models are offered in opposition to the staff's presentation. The staff's strategy of econometric presentation should be one of maintaining an initial focus on primary issues so that some fundamental agreement may be reached prior to the inevitable arguments that concern less significant issues.

The second reason the issue of the technique of presentation of econometric models in regulatory hearings is important lies in the relationship that exists between the use of an econometric model and the problem of procedural due process of law. The relationship between due process and regulatory econometric models will be determined by the attitude of the courts toward a commission's use of facts and associated expertise as formalized in a given model. In its review of regulatory decisions, the Supreme Court has followed the pattern of accepting commission rulings and findings of fact that are supported by substantial evidence. As a method or tool of analysis, the basic econometric approach will undoubtedly be an acceptable commission technique as long as the model has been developed in a manner that does not violate procedural due process of law. It would, however, be difficult to argue that a regulatory commission has observed all of the requirements of due process of law in reaching a given decision if the examiner has relied on an econometric model that has been completely developed by the commission's economics staff without affording an opportunity for cross-examination and rebuttal to all the interested parties. When a complete econometric model is introduced in a hearing, it should be treated as though it is a hostile witness by those who did not have a hand in its development. Interested parties must have the opportunity to evaluate the original development and ultimate impact of each component part of a given econometric model. The development of a viable econometric model to be used as a regulatory device must be accomplished, in part, during the very processes of the scheduled hearing. The model, as a finding of fact, must be based on substantial evidence as presented by parties participating in the hearing—if its use is to conform to the constitutional guarantee of due process.

A set of recommendations was offered in Chapter 5. These recommendations were suggested to improve the basic *Permian* econometric model. It was advised that the Office of Economics strengthen its presentation of new models in future cases by placing greater emphasis on the factors for exploration and supply of natural gas. Consideration should be given to replacing the original dependent variable in the exploratory activities submodel—exploratory wells drilled—with the dependent variable, additions to new reserves. This substitution in format would help bring analytical focus to bear on the desired end product of exploration—the addition of new gas reserves—and not emphasize the drilling of holes. A reformulation of the supply submodel should also give greater emphasis to the variables that bear on decisions to drill primarily wildcat wells since these wells lead to the discovery of new fields. A recommendation has also been made that the Federal Power Commission's staff should engage expert consultants to assist during preparation, presentation, direct examination, and rebuttal testimony in future proceedings.

A final suggestion made in this study is that certain other regulatory agencies might find the econometric approach to ratemaking applicable to their areas of jurisdiction. It has been suggested that the Interstate Commerce Commission, the Civil Aeronautics Board, and the Federal Communications Commission might possibly be able to apply econometric models to aid in the processes of establishing "just and reasonable" rates. It has been suggested that such models might be particularly applicable in cases involving group ratemaking in the case of the Interstate Commerce Commission.

The pioneering efforts of the Federal Power Commission in the investigation and attempted application of an econometric model may provide a stimulus for other commissions to undertake related analyses. The strength of this stimulus will depend, in part, on the degree of success the Federal Power Commission's economics staff achieves in future attempts to employ econometric models.

# Bibliography

**Unpublished Material**

*Analysis of National Exploratory Activity for Hydrocarbons and of Demand for Natural Gas Produced Nationally and in Southern Louisiana.* Federal Power Commission. Docket No. AR 61–2. Exhibit No. 40A.

*Analysis and Projections of Natural Gas Supply and Demand.* Section A and Section B. Federal Power Commission. Docket No. AR 61–1. Exhibits No. 236, 237, and 238.

*Natural Gas: Analysis of National Supply and of National and Southern Louisiana Demand.* Federal Power Commission. Docket No. AR 61–2. Exhibits No. 39 and 40.

*Joint Application for Rehearing of Indicated Respondents, Area Rate Proceeding.* Federal Power Commission. Docket No. AR 61–1.

*Joint Brief on Exceptions of Indicated Respondents in the Matters of Area Rate Proceeding, et al.* Federal Power Commission. Docket No. AR 61–1.

*Joint Brief in Opposition of Indicated Respondents in the Matters of Area Rate Proceeding, et al.* Docket No. AR 61–1.

*Joint Initial Brief of Indicated Respondents, United States of America Before the Federal Power Commission in the Matters of Area Rate Proceedings, et al.* Docket No. AR 61–1. Washington, D.C.

*Joint Reply Brief of Indicated Respondents in the Matters of Area Rate Proceeding, et al.* Docket AR 61–1. January 5, 1964.

*A Layman's Guide to the Wein-Edmonston Econometric Study of Natural Gas Supply and Demand.* March 12, 1964.

*Natural Gas Supply and Demand.* Statement of Harold H. Wein. Federal Power Commission. Docket No. AR 61–1. February 20, 1963.

*Prepared Testimony of J. Harvey Edmonston.* Federal Power Commission. Docket No. 61–2. November, 1964.

*Regulation of the Prices of Natural Gas on an Area Basis: An Analysis of the Permian Basin Opinion.* An address by Bruce R. Merrill before the Independent Natural Gas Association of America. Colorado Springs, Colorado, September 6, 1965.

*Revisions of Certain Pages in Exhibit 235 for 12¢, 17¢, and 22¢ Price Levels to Reflect Changes in Calculations Resulting from Use of Revised Consumer Oil Prices.* Federal Power Commission. Docket No. 61–1. Exhibit 235–C.

*Staff Reply Brief and Response to Examiner's Questions.* Federal Power Commission. Docket No. 61–1. January 15, 1964.

*Staff Reply Brief Opposing Exceptions.* Federal Power Commission. Docket No. AR 61–1. January 15, 1965.

## Books

Adelman, M. S. *The Supply and Price of Natural Gas.* Oxford: Basil Blackwell, 1962.

Baumol, William J. *Economic Theory and Operations Analysis.* Second edition. Englewood Cliffs, N. J.: Prentice-Hall, Inc., 1965.

Brennon, Michael J. *Preface to Econometrics.* Second edition. Cincinnati: South-Western Publishing Company, 1965.

Brown, Robert Goodell *Smoothing Forecasting and Prediction of Discrete Time Series.* Englewood Cliffs, N. J.: Prentice-Hall, Inc., 1963.

Bryant, Edward C. *Statistical Analysis.* Second edition. New York: McGraw-Hill Book Company, 1966.

Commons, John R. *Legal Foundations of Capitalism.* Madison: University of Wisconsin Press, 1959.

Croxton, Frederick E., and Dudley J. Cowden. *Applied General Statistics.* New York: Prentice-Hall, Inc., 1939.

Dixon, Wilfred J., and Frank J. Massey, Jr. *Introduction to Statistical Analysis.* Second edition. New York: McGraw-Hill Book Company, Inc., 1957.

Dorfman, Robert, and others. *Linear Programming and Economic Analysis.* New York; McGraw-Hill Book Company, Inc., 1958.

Duesenberry, James S., and others. *The Brookings Quarterly* Econometric Model of the United States. Chicago: Rand McNally & Company, 1965.

Fisher, Franklin M., *Supply and Costs in the U.S. Petroleum Industry.* Baltimore: Johns Hopkins Press, 1964.

Fox, Karl A. *The Theory of Quantitative Economic Policy with Applications to Economic Growth and Stabilization.* Vol. 5. Chicago: Rand McNally & Company, 1966.

Freund, John E. *Mathematical Studies.* Englewood Cliffs, N. J.: Prentice-Hall, Inc., 1962.

Friedmann, W. *Law in a Changing Society.* Baltimore: Penguin Books Ltd., 1964.

Hadley, G. *Introduction to Probability and Statistical Decision Theory.* San Francisco: Holden-Day, Inc., 1967.

Hawkins, Clark A. *The Field Price Regulation of Natural Gas.* Tallahassee: Florida State University Press, 1969.

Hodges, John E. and Henry B. Steele. *An Investigation of the Problems of Cost Determination for the Discovery, Development, and Production of Liquid Hydrocarbon and Natural Resources.* Vol. XLVI, No. 3. Houston, Texas: The Rice Institute, 1959.

Huang, David S. *Introduction to the Use of Mathematics in Economics Analysis.* New York: John Wiley & Sons, Inc., 1964.

Kaufman, Arnold. *Methods and Models of Operations Research.* Englewood Cliffs, N. J.: Prentice-Hall, Inc., 1964.

Klein, Lawrence R. *An Introduction to Econometrics.* Englewood Cliffs, N. J.: Prentice-Hall, Inc., 1962.

Liebhafsky, H. H. *The Nature of Price Theory.* Homewood, Illinois: The Dorsey Press, Inc., 1963.

Malinvaud, E. *Statistical Methods of Econometrics.* Chicago: Rand McNally & Company, 1966.

Papandreau, Andreas G. *Competition and Its Regulation.* New York: Prentice-Hall, Inc., 1954.

Phillips, Charles F., Jr. *The Economics of Regulation.* Homewood, Illinois: Richard D. Irwin, Inc., 1965.

Richmond, Samuel B. *Statistical Analysis.* Second edition. New York: The Ronald Press Company, 1964.

Samuelson, Paul Anthony. *Foundations of Economic Analysis.* New York: Anthenum 1965.

Stockton, John R., and others. *Economics of Natural Gas In Texas.* Austin, Texas: Bureau of Business Research, College of Business Administration, The University of Texas, 1952.

Theil, Henri. *Applied Economic Forecasting.* Chicago: Rand McNally & Company, 1966.

Theil, Henri, John C. G. Boot, and Teun Klock. *Operations Research and Quantitative Economics.* New York: McGraw-Hill Book Company, 1965.

Walters, A A. *An Introduction to Econometrics.* New York: W. W. Norton Company Inc., 1970.

Wilcox, Clair. *Public Policies Toward Business.* Chicago: Richard D. Irwin, Inc., 1955.

## Articles and Periodicals

Colwell, B. Joe. "Natural Gas Area Pricing: Economic and Legal Considerations," *Southwestern Social Science Quarterly* 48, no. 2 (September 1967).

*New York Times.* 1959–1964.

Samuelson, Paul A., and others. "Report of Evaluation Committee of Econometrics," *Econometrics* 22, no. 2 (April 1954).

Theil, H. "Econometrics and Management Sciences: Their Overlap and Interaction," *Management Science* 11, no. 8 (June 1965).

## Other Sources

American Gas Association. *Proved Reserves of Crude Oil, Natural Gas Liquids and Natural Gas.* New York: American Gas Association, 1964.

Federal Power Commission. *Hearing Examiner's Initial Opinion.* Docket No. AR 61–1. 1965.

Federal Power Commission. *Presiding Examiner's Initial Decision on Permian Area Rates.* Docket No. AR 61–1. Washington: Government Printing Office, 1964.

*Federal Register.* National Archives of the United States. Vol. 5, No. 139 (July 20, 1950).

*Final Report of the United States Fuel Administration 1917–19.* Washington: Government Printing Office, 1921.

McKie, James W. *The Regulation of Natural Gas.* Washington: The American Enterprise Association, June 1957.

*Opinion and Order Determining Just and Reasonable Rates for Natural Gas Producers in the Permian Basin.* Federal Power Commission. Docket No. AR 61–1. Opinion No. 468. August 5, 1965.

*Presiding Examiner's Initial Decision on Southern Louisiana Area Rates.* Federal Power Commission, Docket No. AR 61–2. December 30, 1966.

*United States Government Organization Manual,* 1967–68. Office of the Federal Register, National Archives and Records Service, General Services Administration. Washington, D.C., 1967.

# Notes

## Foreword

1. For a discussion of this point of view in the area of political economy, see my *American Government and Business* (New York: John Wiley & Sons, 1971), pp. 23ff.; pp. 37ff.; and pp. 569ff.

2. See *Denver and Rio Grande Western Railroad, et al.* v. *U.S.*, 387 U.S. 486 (1967).

## Chapter 1

1. Harold H. Wein, *Natural Gas Supply and Demand* (Washington, D.C.: Federal Power Commission, Office of Economics, Docket No. AR 61-1, 1963), p.6.

## Chapter 2

1. National Archives and Records Service, General Services Administration, *United States Government Organization Manual 1967–68* (Washington: U.S. Government Printing Office, 1968), p. 450.

2. *Final Report of the United States Fuel Administrator 1917–19* (Washington: U.S. Government Printing Office, 1921), pp. 39–40.

3. *Missouri* v. *Kansas Natural Gas Co.*, 265 U.S. 298 (1924).

4. *Utility Corporations: Final Report of the Federal Trade Commission on Economic, Corporate, Operating, and Financial Phases of the Natural-Gas-Producing, Pipe-line and Utility Industries* (Senate Document 92, Part 84A, Chapter 13), pp. 607–617.

5. *H. R. Report, No.* 709, 75th Congress, First Session, pp. 1–2.

6. *Ibid.*, p. 20.

7. John R. Stockton, R. C. Henshaw, R. W. Graves, *Economics of Natural Gas in Texas* (Austin, Texas: Bureau of Business Research Monograph No. 15, 1952), p. 219.

8. United States Code., sub-section 717 (f).

9. *Peoples Natural Gas Company* v. *Federal Power Commission*, 217 Fed (2d) 153 (1942), Certiorari denied 316 U.S. 700 (1942).

10. *Interstate Natural Gas Company* v. *Federal Power Commission*, 331 U.S. 682 (1947).

11. H. Doc. No. 555, *Congressional Record*, vol. 96 (April 18, 1950), pp. 5304–5305.

12. *Federal Register*, National Archives of the United States, vol. 15, no. 139 (July 20, 1950), p. 4633.

13. *Phillips Petroleum Company* v. *Wisconsin*, 347 U.S. 672 (1954).

14. Federal Power Commission Opinion No. 246 (1951).

15. *State of Wisconsin et al.* v. *Federal Power Commission*, 92 U.S. App. D.C. 284, 205 Fed (2d) 706 (1953).

16. *Phillips Petroleum Company* v. *Wisconsin*, 347 U.S. 672 (1954).

17. *New York Times*, 10:6 (February 3, 1956).

18. *Congressional Record*, 84th Congress, Second Session (February 17, 1956), p. 2793.

19. Charles F. Phillips, Jr., *The Economics of Regulation* (Homewood, Illinois: Irwin, 1965), pp. 593–594.

20. M. S. Adelman, *The Supply and Price of Natural Gas* (Oxford: Basil Blackwell, 1962), p. 28.

21. B. Joe Colwell, "Natural Gas Area Pricing: Economic and Legal Considerations," *Southwestern Social Science Quarterly*, Vol. 48, No. 2 (September, 1967), p. 201.

22. Federal Power Commission, Opinion No. 468, Docket No. AR 61–1, (August 5, 1965), pp. 7–8.

23. *Bel Oil* v. *Federal Power Commission*, 255 Fed. (2d) 548 (1958), Certiorari denied 358 U.S. 804 (1958).

24. Federal Power Commission Opinion No. 468, (1965), pp. 9–10.

25. *Ibid.*

26. *Wisconsin* v. *Federal Power Commission*, 373 U.S. 294 (1963).

27. *In Re Permian Basin Area Rate Cases*, 88 S. Ct. 1344 (1968).

## Chapter 3

1. Paul Anthony Samuelson and others, "Report of Evaluation Committee on Econometrics," *Econometrics* 22, no. 2 (April 1954): 142.

2. K. A. Fox, J. K. Sengupta, and E. Thorbecke, *The Theory of Quantitative Economic Policy* (Chicago: Rand McNally, 1966), p. 4.

3. James S. Duesenberry, Gary Fromm, Lawrence R. Klein, and Edwin Keeh (eds.), *The Brookings Quarterly Econometric Model of the United States* (Chicago: Rand McNally, 1965), p. 14.

4. *Pennsylvania* v. *West Virginia* (Dissenting Opinion), 262 U.S. 553, 621 (1923).

5. *Colorado Interstate Gas Company* v. *Federal Power Commission*, 324 U.S. 581, 612 (1945).

6. H. Theil, *Optimal Decision Rules for Government and Industry* (Chicago: Rand McNally, 1964), p. 124.

7. Harold H. Wein, *National Gas Supply and Demand* (Washington: D.C.: Federal Power Commission, February 20, 1963), p. 33.

8. *Ibid.*, p. 35.

9. *Ibid.*, p. 10.

10. "Joint Initial Brief of Indicated Respondents," *United States of America before the Federal Power Commission in the Matters of Area Rate Proceedings, et. al.*, Docket No. AR 61–1 (Washington, D.C.), pp. 23–24.

11. Federal Power Commission, *Hearing Examiners Initial Opinion*, Docket No. AR 61–1 (1965), p. 33.

## Chapter 4

1. E. Malinvaud, *Statistical Methods of Econometrics* (Chicago: Rand McNally, 1966), p. 49.

2. Franklin M. Fisher, *Supply and Costs in the U.S. Petroleum Industry* (Baltimore: John Hopkins Press, 1964), p. 12.

3. Harold H. Wein, *Natural Gas Supply and Demand* (Washington, D.C.: Federal Power Commission, Office of Economics), Docket No. AR 61–1, 1963), p. 6.

4. *Ibid.*, p. 43.

5. *Ibid.*, p. 44.

6. *Ibid.*, p. 50.

7. *Ibid.*, p. 57.

8. *Ibid.*, p. 82.

9. American Gas Association, *Proved Reserves of Crude Oil, Natural Gas Liquids and Natural Gas* (New York: American Gas Association, 1964), p. 23.

10. Wein, *op. cit.*, p. 13.

11. *Joint Reply Brief of Indicated Respondents, in the Matters of Area Rate Proceeding, et al.,* Docket AR 61–1 (January 15, 1964), pp. 46–47.

12. *Joint Initial Brief of Indicated Respondents, in the Matters of Area Rate Proceeding, et. al.,* Docket No. AR 61–1, pp. 69–70.

13. *Ibid.,* p. 19.

14. Lawrence R. Klein, *An Introduction to Econometrics* (New Jersey: Prentice Hall, 1962), pp. 62–64.

15. *Joint Reply Brief, op. cit.,* p. 60.

16. *Joint Initial Brief, op. cit.,* p. 26.

17. "Opinion and Order Determining Just and Reasonable Rates for Natural Gas Producers in the Permian Basin," Opinion No. 468 (August 5, 1965), p. 28.

18. *Joint Reply Brief, op. cit.,* pp. 56–57.

## Chapter 5

1. 30 Federal Power Commission 1356.

2. J. Harvey Edmonston, *Prepared Testimony Southern Louisiana Area Rate Proceeding,* Docket No. AR 61–2, p. 1745.

3. Harold H. Wein, *Natural Gas Supply and Demand* (Washington, D.C.: Federal Power Commission, Office of Economics) Docket No. AR 61–1, 1963, p. 77.

4. Federal Power Commission, *South Louisiana Area Exhibit* 252, Docket No. AR 61–2, p. 30, 016.

5. Federal Power Commission, *Southern Louisiana Area Exhibit* 170, Docket No. AR 61–2, p. 22, 500.

6. *Amerada Reply Brief,* Southern Louisiana Area Rate Proceeding, Docket No. AR 61–2, pp. 21–22.

7. *Ibid.,* pp. 23–29.

8. *Ibid.,* p. 30.

9. J. Harvey Edmonston, *op cit.,* p. 1747.

10. Federal Power Commission, *Exhibit* 131, Docket No. AR 61–2, p. 18,082.

11. Federal Power Commission, *Southern Louisiana Area Rate Proceeding Exhibit* 171, Docket No. AR 61–2, p. 22, 220.

12. *Presiding Examiner's Initial Decision on Southern Louisiana Area Rates*, Federal Power Commission, Docket No. AR 61–2 (December 1966), p. 232.

13. Federal Power Commission, *Staff Initial Brief*, Docket No. AR 61–2, p. 334.

14. *Presiding Examiner's Initial Decision on Southern Louisiana Area Rates*, *op. cit.*, pp. 232–233.

15. *Ibid.*, p. 254.

16. *Presiding Examiner's Initial Decision on Permian Area Rates*, (September 17, 1964), p. 33.

17. Robert Goodell Brown, *Smoothing Forecasting and Prediction of Discrete Time Series* (Englewood Cliffs, New Jersey: Prentice Hall, 1963), p. 77.

18. *Ibid.*, p. 221.

## Chapter 6

1. Federal Power Commission, *Annual Report* (Washington: U.S. Government Printing Office, 1962), p. 103.

## Chapter 7

1. John R. Commons, *Legal Foundations of Capitalism* (Wisconsin: University of Wisconsin Press, 1959), p. 331.

2. Justice Roberts, *Nebbia* v. *New York*, 291 U.S. 502,525 (1934).

3. John R. Commons, *op. cit.*, p. 329.

4. United States, *Statutes at Large*, Volume 60 Part I (1947).

5. *Ibid.*, p. 240.

6. Paul J. Garfield and Wallace F. Lovejoy, *Public Utility Economics* (Englewood Cliffs, New Jersey: Prentice-Hall, 1964), pp. 41–42.

7. *Interstate Commerce Commission* v. *Illinois Central R. R. Co.*, 215 U.S. 452 (1960), p. 470.

8. *Federal Power Commission* v. *Natural Gas Pipeline Co.*, 62 Supreme Court 736 (1942), p. 742.

9. *Penn-Central Merger and N and W Inclusion Cases*, 88 Supreme Court 602 (1968).

10. *Ibid.*, p. 601.

11. *Ibid.*, pp. 621–622.

12. *Federal Power Commission* v. *Hope Natural Gas Co.*, 320 U.S. 602 (1944).

13. *In Re Permian Basin Area Rate Cases*, 88 Supreme Court 1344 (1968), p. 1360.

**Chapter 8**

1. *Phillips Petroleum Company* v. *Wisconsin*, 347 U.S. 677 (1954).

# Index